Research on Innovation of Express Delivery Network Management

Pengfei Li · Jianhong Wu

Research on Innovation of Express Delivery Network Management

Network Optimization, Operation Mechanism and Mode Reform

SSAP

Springer

Pengfei Li
Xi'an University of Posts
and Telecommunications
Xi'an, China

Jianhong Wu
Xi'an University of Posts
and Telecommunications
Xi'an, China

Translated by
Rong Chen
Xi'an University of Posts
and Telecommunications
Xi'an, China

Yanyan Tian
Xi'an University of Posts
and Telecommunications
Xi'an, China

Xinyu Zhou
Xi'an University of Posts
and Telecommunications
Xi'an, China

Zhixiong Fang
Xi'an University of Posts
and Telecommunications
Xi'an, China

ISBN 978-981-99-8981-2 ISBN 978-981-99-8982-9 (eBook)
https://doi.org/10.1007/978-981-99-8982-9

Jointly published with Social Sciences Academic Press
The print edition is not for sale in The Mainland of China. Customers from The Mainland of China please order the print book from: Social Sciences Academic Press

Translation from the Chinese Simplified language edition: “快递网络管理创新研究” by Pengfei Li et al., © Social Science Academic Press 2020. Published by Social Science Academic Press. All Rights Reserved.

This Springer imprint is published by the registered company Springer Nature Singapore Pte Ltd.
The registered company address is: 152 Beach Road, #21-01/04 Gateway East, Singapore 189721, Singapore

Preface

The escalating pace of Internet and e-commerce advancements has engendered a surge in consumerism, fostering a favorable outlook on purchasing behavior. Consequently, the need for expedited delivery services has witnessed a notable upswing. Express delivery plays a pivotal role in facilitating social and economic advancement, serving as a substantial contributor to the national economy while yielding substantial benefits to individuals. In order to foster the industry's growth, it is imperative to fortify the infrastructure of the express delivery network and drive innovation in its managerial approach. The pertinent research endeavors revolve around network optimization, operational mechanisms, and mode transformation, with the aim of comprehending the operational intricacies of the network and identifying avenues for innovative mode transformation. Consequently, a comprehensive and novel framework for effective management can be established. This research endeavors to enhance the quality and efficiency of express services through upgraded levels of management and optimized industry practices.

This book has been developed with a keen understanding of the practical requirements of express companies and is tailored to encompass the unique characteristics of express delivery networks. It boasts the following four prominent features:

1. An evaluative index system for assessing the progress of the industry has been meticulously devised. By analyzing a range of factors, 13 indicators have been identified, encompassing the macroeconomic environment, system investments, industry growth, and the degree of informatization. These indicators collectively form the foundation of the system. Employing factor analysis, key influential factors are extracted, including fundamental economic factors, input element factors, and factors pertaining to industrial development.
2. A model for optimizing the path of express delivery networks is formulated, utilizing intelligent algorithms. In order to tackle the intricate nature of express delivery networks, two intelligent algorithms, namely the ant colony algorithm and the simulated annealing algorithm, are incorporated for optimization

purposes. Through empirical research conducted on these two optimization algorithms, it is demonstrated that they effectively address the challenge of network path optimization.
3. A novel approach for assessing the operational efficiency of express delivery networks is introduced, employing a three-stage Data Envelopment Analysis (DEA) method to analyze the operational efficiency and overall factor productivity of the nationwide express delivery network. The empirical analysis results demonstrate that this method effectively captures the dynamic fluctuations in the development status of express delivery networks across 31 provinces in China. Furthermore, it enables the identification of factors that influence their operational efficiency, thereby enhancing the evaluation system for express operations' efficiency.
4. The management system of express delivery networks undergoes significant innovation. Through an examination of the management system of the existing national express delivery networks in China and an analysis of deficiencies from five key aspects (innovative ideas, approaches, infrastructure construction, technology operating costs, and talent training), the authors suggest the construction of an innovative ecosystem for the national express delivery network management system. This entails reforming management modes and conceptual frameworks to address the identified shortcomings.

This pioneering book serves as the inaugural specialized resource offering a comprehensive analysis of research on express delivery networks management. By engaging with its contents, readers can acquire knowledge in the following five key areas:

1. The evolution of express companies and the state of express delivery networks construction;
2. The structure and characteristics of express delivery networks;
3. Decision-making methods and evaluation systems for assessing the level of development in the optimization of express delivery networks;
4. The operation mechanism of express delivery networks and the efficiency of their operation;
5. The management mode and innovative management system of express delivery networks.

This monograph is well-suited for a diverse range of readers, including postgraduate students, scholars, planning practitioners, decision-makers, and logistics enterprise managers. It serves as a valuable resource for research and reference within the fields of logistics engineering, supply chain management, and transport planning and management. By delving into the content of this book, individuals within these domains can gain valuable insights, expand their knowledge base, and utilize the information for academic pursuits, practical applications, and informed decision-making processes.

Xi'an, China

Pengfei Li
Jianhong Wu

Contents

Chapter 1
Introduction

1.1 Background and Significance

The Document No. 1 issued by the General Office of the State Council in 2018, entitled "Opinions on Promoting the Coordinated Development of E-commerce and Express Delivery Logistics," emphasized the importance of the e-commerce express industry. In the Report on the Work of the Government, Premier Li Keqiang stated that it is necessary to "encourage the rapid growth of express delivery and online commerce." In 2014, the express delivery industry was mentioned for the first time in the Report on the Work of the Government. Since then, it has been included for five consecutive years. In 2017, the national service sector's value added to GDP reached 51.6%, and the share of final consumption expenditures to GDP grew to 58.8%. The tertiary industry has become an increasingly important pillar of economic growth. The macroeconomic environment has aided the logistics and express delivery sector in achieving its present momentum. However, the growth of the express delivery business is likely to slow down and eventually reach maturity as the development of e-commerce approaches its "limit" of growth in the coming years. In order to exceed the growth cap, the future growth must be reliant on industrial upgrading and transformation. Consequently, it is necessary to conduct research in the direction of the evolution of the express delivery and logistics industry. The supply-side structural reform of the logistics and express delivery industries will be expedited while being guided by the principle of scientific development.

The express delivery industry has experienced rapid growth in conjunction with the unprecedented expansion of the e-commerce sector, and has had a major impact on many aspects of the daily lives of local residents and the development of the national economy. By the end of 2017, the total national logistics volume had reached 47.94 billion tons, with a cargo turnover of 19,737.3 billion ton-kilometers. Of these, road freight amounted to 36.80 billion tons, with a cargo turnover of 6,671.25 billion ton-kilometers; railway freight was 3.69 billion tons, with a cargo turnover of 2969.22 billion ton-kilometers; and civil aviation cargo totalled 24.35 billion ton-kilometers.

P. Li and J. Wu, *Research on Innovation of Express Delivery Network Management*,
https://doi.org/10.1007/978-981-99-8982-9_1

The majority of transportation tasks were shared by the network of roads, railroads, and waterways, which served as the primary transportation network for logistics express delivery. The highway network, which handled 76.8% of the transportation responsibilities amongst them, was crucial. In 2017, the volume of express delivery business reached 40.06 billion pieces, representing a year-on-year increase of 28%, and the load on the express delivery network continued to increase. Nevertheless, the express delivery business must address and find solutions to issues such as increasing logistics and transportation efficiency through creating layout, planning routes, managing resources efficiently, reducing operating costs, and improving service quality.

Therefore, the management of express delivery networks is particularly important, and new technologies such as big data and cloud computing can provide a new perspective for assessment. The implementation of new technology-based management innovation on express delivery networks has the following practical implications.

1. Express delivery enterprises can benefit from forecasting, identifying, and resolving various challenges that need to be addressed in the construction of logistics network management.
2. It enhances the operational efficiency of the express delivery industry by providing managerial strategies and solutions for companies and their managers.
3. It improves the quality of service in the express delivery sector while decreasing operational costs.

Consequently, the innovation of express delivery networks management is a crucial issue that must be addressed by express delivery enterprises in their course of development.

1.2 Literature Review

Research on the express delivery networks management currently focuses on the following aspects.

1.2.1 Optimization of Express Delivery Networks

1.2.1.1 Structure of Express Delivery Networks

(1) The Structure of Express Delivery Networks

Chestler (1985) found that a hub-and-spoke network contains hubs and conducted research on the location of these hubs. O'Kelly (1987) was the first to propose the use of Binary Integer Programming to select the hub location for this type of network. Subsequently, a number of scholars have undertaken research into the models and algorithms of Hub-and-spoke networks, which can be divided into two categories. The first group focuses on improving the binary integer programming model proposed by Horner and O'Kelly (2001) discovered that the transportation discount rate between hubs increases with the rising traffic volume; Pels et al. (2000) performed a comparative analysis of the two types of network structure under the condition of parameter alteration; Brueckner and Zhang (2001) examined network topology using the Hub-and-spoke and fully connected approaches while considering flight frequency and airline cost. Bai and Zhu (2006) studied the flow of passengers, flight frequency, and number of seats on a single aircraft within both the fully connected airline network and the hub airline network. Cunha and Silva (2007) suggested that unit price of transportation between nodes and hubs decreases with increased transportation volume and that each edge in the hub-and-spoke network reflects economies of scale. Jeong et al. (2007) highlighted that express delivery networks can be divided into hub-and-spoke networks with and without capability limits, depending on whether the hub has capacity constraints or not. Furthermore, the impact of traffic congestion on the hub was also taken into consideration.

The second group further investigates the Hub-and-spoke express delivery network structure; Lin and Lin (2007) conducted a study on the design of the Hub-and-spoke network structure under random demand. De Camargo et al. (2009) classified express networks into single-hub networks and multi-hub networks based on the number of hubs. Moreover, the hub network has been identified in the railway freight networks of France, Italy, Germany, and 10 other European countries; Andreas et al. (2009) studied single-distribution and multi-distribution uncapacitated hubs, and provided an integer programming formula for the problem. Additionally, he proposed the branching delimitation method to address multi-distribution. Yang et al. (2016) designed an optimization algorithm of express delivery network based on complex network theory. This algorithm gradually deleted the edges with the lowest marginal benefit from the fully connected network, thus iteratively finding the network structure with the lowest connection cost. The effectiveness of the algorithm was verified through examples.

Previous studies have expanded the notion of express delivery networks from solely interconnected networks or hub-and-spoke networks to hybrid networks. Scholars have conducted a detailed exploration of the economic implications involved, which has enriched the theories on the structure of express delivery networks. Despite the lack of adequate consideration of the practical conditions of operations, such as the timeliness of express delivery services and financial constraints, it is necessary to explore the implications of express delivery network structure based on increasingly more mature network structure theories, in order to improve the express delivery network structure and make the optimization of express delivery networks more feasible.

(2) The Structural Characteristics of Express Delivery Networks

The structural characteristics of express delivery networks form the basis of optimizing the structure and selecting the design method. Armacost et al. (2004) analyzed the express delivery network of UPS and identified that it possessed partition characteristics. Furthermore, nodes exist in all regions that are responsible for the transmission and reception of commodities in given areas. As a result of various operational modes, different express delivery network designs, such as the hub-and-spoke or fully-linked network structures, have emerged. Their characteristics and differences have then been further explained. For instance, Smirti et al. (2007) evaluated the pyramid network structure of FedEx, considered the cost of sorting and transportation individually, and compared the effect of each additional level on the overall express delivery networks and the alteration in operating costs.

Yang et al. (2009) examined the characteristics of express delivery networks topology from the perspective of complex networks, finding that it exhibited a small-world or rich people's club phenomenon, characterized by heterogeneity and mixed behavior. Huang (2009) constructed a trunk line of express delivery enterprises and conducted an analysis of the intricate transportation network diagram, which revealed the network to be scale-free and small-world in nature. It was observed to be robust under random interference, yet incredibly fragile when subject to deliberate damage. It was further suggested that nodes with a greater degree value should be closely monitored due to their potential to have major effects on the networks.

Zhang and Nie (2009) discovered that express delivery networks can be conceptualized as an additional type of transportation networks, with each network node functioning as a node and the transportation route as boundary. The network is structured in a hybrid model with three layers: arterial roads within a city, within a region and between regions, all connected by urban thoroughfares.

The current research on the structural characteristics of express delivery networks can be summarized into two aspects: empirical study of the express delivery network structure, and exploration of the topological characteristics from the perspective of complex networks. This has laid a theoretical foundation for express delivery networks planning and design. Nevertheless, discrepancies between these findings and the operation of express delivery companies in real life still exist, and require improvement and optimization. Further subdivision and in-depth research on the characteristics of the backbone network and node characteristics of express delivery networks is also necessary.

1.2.1.2 Express Delivery Network Planning and Design

Network planning and design are essential for optimizing delivery networks, yet they can be challenging (Armacost et al., 2002). Barnhart and Schneur (1996) used air express as their research object and created a model of the express delivery network with only one hub. Additionally, a radial network structure was chosen to incorporate logistics resources. The research indicated that the Hub-and-spoke network structure is commonly used for planning and designing road transportation networks and hybrid networks that combine road and aviation (Kuby & Gray, 1993; Wasner & Zaepfel, 2004).

However, China's express delivery industry is still in the initial stage of development. Zhang et al. (2008) compared a composite Hub-and-Spoke network with direct routes to a pure Hub-and-Spoke transportation network based on a highway express delivery network, and concluded that the former has a significant impact on decreasing the number of goods transshipments, shortening the service time and reducing operating costs. Dai and Huo (2010) analyzed the planning of non-hub-spoke express delivery networks for multiple cargo flows. Li and Ding (2012) selected a hub-and-spoke network to plan and design hub locations and distributions in order to reduce operational costs, meet timeliness requirements for transit goods, and improve delivery efficiency. A multi-objective hub-and-spoke express delivery network model was constructed by standardizing the fitness function with the help of a simulated annealing algorithm. Zhao et al. (2016) proposed a planning decision-making model of hybrid hub-and-spoke express delivery network that allows direct access from a global optimization perspective when the express delivery business volume reaches a certain scale, eliminating the need for a transit center. This model reduces the total cost of the network, mitigates detours, and improves service timeliness and quality on the direct route. Lu (2016) conducted an analysis of road, railway, and air transportation by constructing mathematical models from two perspectives: cost penalty and time threshold, and comparing and analyzing the results.

1.2.1.3 Network and Path Optimization Methods

Currently, two main types of path and network optimization methods exist: accurate algorithms (Dijkstra, 1959; Jiang et al., 2014; Floyd, 1962; Ren & Yang, 2007; Fisher, 1994; Padberg & Rinaldi, 1991; Kohl & Madsen, 1997) and soft computing methods (Tan et al., 2006; Zhao et al., 2004; Li & Zhang, 2006; Yan & Wu, 2009; Xu et al., 2005; Zhang et al., 2009; Li et al., 2004; Fu, 2009). A comparative analysis of the typical algorithms in these two types of methods is presented in Table 1.1.

Karkazis and Boffey (1995) employed the Branch and Bound method for optimizing the route of dangerous goods in real life and suggested considering uncertain factors such as weather and random population distribution in future research. Kulak et al. (2012) implemented a Cluster-based Tabu Search Algorithm for multi-batch distribution path optimization, which has the advantage of adjustable parameters. Marie et al. (2012) employed the ant colony algorithm to study the shortest path problem of multi-modal transport, and found that it is suitable for solving large-scale problems. Braekers et al. (2012) studied the container path optimization problem of multi-modal transport terminals. Assuming that the origin or destination was unknown, taking quantity and distance as the goals, a dual-objective programming model with time windows was designed, and simulated annealing algorithm and tabu search algorithm were selected for optimization, respectively.

Currently, there is an increasing number of algorithms proposed for research on network and path optimization. However, the majority of these only consider a single objective, while express delivery optimization involves multiple factors such as

Table 1.1 Comparison of path and network optimization methods

Category of Method	Algorithm	Pros and cons
Accurate Algorithm	Dijkstra's Algorithm	These algorithms have a high degree of computational accuracy, but they require a long time to calculate, and their complexity of solution is high, making them unsuitable for use in large-scale networks.
	SAPR Algorithm	
	Floyd Algorithm	
	Linear Programming	
	K-tree Algorithm	
	Method of Branch and Cut	
	Lagrange Relaxation Algorithm	
Soft Computing Methods	Evolutionary Algorithm(EA)	Soft computing methods that involve algorithms possess significant parallel operation characteristics and are suitable for solving network path optimization problems with multiple objective functions, more influential factors, and higher complexity.
	Ant Colony Algorithm	
	Genetic Algorithms	
	Particle Swarm Algorithm	
	ADD Algorithm	
	Immune Algorithms	
	Predatory Search Algorithm	
	Fuzzy Theory	

distance, time, and environment. Thus, a single objective is insufficient to adequately address practical issues.

The earliest research conducted by Chinese scholars focused on the China Postal Network. Hu and Zhang (2001) identified the challenges existing in the backbone network of China Post and proposed a multi-objective hierarchical decision-making model, which was composed of three components: train optimization, route optimization and delivery optimization. Ye et al. (2004) employed the Lagrangian relaxation method and column generation method to address the inter-provincial transshipment problem, transforming it into a network planning problem with time and capacity constraints. Fan and Wang (2006) applied an immune algorithm to solving the vehicle routing problem, resulting in a range of solutions while avoiding convergence to a locally optimal solution, thereby successfully avoiding the risk of losing the optimal solution during the evolutionary process.

With the rapid growth of the express delivery industry, it was only in 2007 that Chinese scholars began to apply the hub-and-spoke structure theory to express delivery network optimization. Zhang et al. (2008) then applied a hybrid shaft-spoke structure theory to highway freight planning, constructing a hybrid shaft-spoke network with direct routes, and utilizing a genetic algorithm to find solutions. Yang and Zou (2011) developed an optimization model to design a highway passengers express delivery network in order to reduce express transportation time and enhance the transportation capacity. Ge et al. (2013) constructed a mixed-integer programming hub-and-spoke network model with the time threshold and the tariff discount factors of trunk and branch transportation as constraints, with the aim of optimizing the regional express network. Yang et al. (2015) employed the Floyd algorithm and Dijkstra algorithm to construct an express delivery network optimization model with delivery time and node maximum flow constraints, taking into account the delivery cost of express networks. He et al. (2019) optimized the "last mile" delivery route planning problem for express delivery companies by utilizing an improved tabu search algorithm. This algorithm yielded optimized route results quickly, and was able to dynamically process new demands and real-time requirements from delivery activities, thereby enhancing delivery efficiency by updating the route.

Currently, the objective function handling in network and path optimization is inadequate. Generally, the existing objective functions only factor in the transportation cost. While some scholars have taken into account the waiting cost, they have largely simplified this treatment. Furthermore, the time threshold of goods at nodes and the transportation cost of different road sections are both related to the operational efficiency of the entire network, and the environmental cost is a factor that should be given due consideration.

1.2.2 Operation Mechanism of Express Delivery Networks

The operational mechanism of express delivery networks is a generalization of the structure and mode of operation between the internal and external factors of the business activities of express delivery companies. Low operational efficiency is a major limiting factor for the development of China's postal industry. Xu et al. (2007) investigated the current state of regional logistics and concluded that the most effective way to improve the operational efficiency of logistics services is to implement informatization and networking of logistics, as well as to conduct thorough research on the principles, participants, and benefit realization modes of the operational mechanism. An efficient and harmonious regional logistics information network system should be established to meet the high-level needs of logistics informatization, as suggested by Wang et al. (2016). Through the use of a two-stage DEA model, an empirical study was conducted to evaluate the operational efficiency, market efficiency, and overall operational efficiency of China's postal industry. The results indicated that the overall operation efficiency of postal service in China is low, with significant regional disparities, and that the market efficiency was higher than the operation efficiency.

Operational efficiency directly reflects the quality of the operational mechanism, while the latter determines the level of operational efficiency. Li and Yang (2018) investigated ways to leverage the network platform operational mechanism to acquire knowledge from other participants in the platform and enhance NSD (Internet+shop+direct sell) performance, thereby constructing a theoretical model illustrating how the network platform operational mechanism impacts content providers' NSD performance. The findings showed that content providers can improve operational efficiency through capability provision mechanism, interaction mechanism, and knowledge integration mechanism.

Consequently, many scholars have conducted research and analysis on the logistics operation mechanism in order to attempt to improve operational efficiency. Li et al. (2014) conducted an on-site investigation and found that the current bonded logistics system is lacking coordination in its management system, and has a poor operation mechanism with "two ends outside" and an unclear industrial service concept. Furthermore, the system lacks support and driving forces from demands, with advanced construction existing alongside neglected hardware facilities. Additionally, there are various problems such as insufficient transportation network construction and low efficiency of collection and distribution operations between the hinterland and the market. Thus, it is imperative to establish a unified coordination mechanism. Zhang (2015) took the supply chain operating system for agricultural product logistics as the object of his research. He then discussed the coordination elements, logistics composition, operation mechanisms, and profitability of the agricultural product supply chain, based on the supply chain coordination and related theories of agrarian product supply chain, with the aim of exerting a positive impact on the agricultural product logistics operation mechanism in the future. Li (2017) believed that the operation mechanism of a modern logistics park determines the orientation and positioning of its development. Wu et al. (2007) provided an overview of the unique regional external environment for the development of cold chain logistics for aquatic products in Fujian Province. Subsequently, employing the SCP model, the authors conducted an analysis of the structure of the industry, market behavior, and current market performance of the cold chain logistics operation mechanism for aquatic products in Fujian Province. In the end, they proposed a model to improve the cold chain logistics operation mechanism for aquatic products in Fujian Province, which included the construction of an integrated industrial information platform, the development of a regional sharing economy, and the optimization of the spatial layout of the low-temperature logistics distribution center.

Few relative studies exist on the management and operational mechanisms of express delivery networks, let alone a mature and comprehensive theoretical system. As such, the authors also refer to the operational mechanism studies in other fields for this research.

Ding et al. (2008) discussed the major issues associated with current higher vocational college-enterprise cooperation from an operational perspective. By taking into consideration the prevailing circumstances, they proposed a dynamic, benefit-driven, guarantee, sharing, and communication mechanism to innovate approaches to such cooperation. Xiang et al. (2016) explored the accelerator's operational mechanism in the context of the "mass entrepreneurship and innovation" movement, and thoroughly elucidated the four fundamental operational mechanisms of accelerators, namely the access, service, profit, and exit mechanisms. Through an in-depth analysis of foreign accelerators as well as the prevailing accelerator operational model in

China, countermeasures and advice for the operation and construction of accelerators were then suggested. Tan and Xu (2016) investigated the effective strategies that financial companies can use to link with their head offices in practice. By designing a performance evaluation index system, a business authorization review mechanism, and an operating income distribution mechanism, they argued that it is essential to establish a financial company operation control mechanism that is based on the integration of interests.

Wang and He (2008) proposed the internal mechanism of operation of mutual funds through an analysis of the relationship between the community game and the mutual fund game. Zhao (2004) employed Social Network Analysis (SNA) to construct a network model of resource flow within the alliance and proposed an alliance operation mechanism based on digital publishing platform communication and online shopping platform transaction for cost and risk reduction.

A considerable number of scholars have made significant contributions to the development of operational mechanisms. Feng and He (2010) argued that, in the knowledge and economic era, organizations primarily attain sustainable competitive advantage through knowledge and the capability to generate and utilize knowledge. After analyzing the organizational knowledge innovation system and its operational mechanism, it was determined that its effective operational mechanism comprises of the decision-making mechanism, the dynamic mechanism and the assurance mechanism. Wu and Lin (2010) conducted a further study of the operational status and bottleneck factors of the national logistics public information platform, based on an analysis of existing literature. As a solution, they proposed a "coordinated" operation between operators and the government to address the operational shortcomings. Building on the innovative practice of SMG operational mechanisms, Lin (2017) explained the mechanism in terms of its driving force, ecological transformation, and operational innovation, suggesting methods for preserving the dynamism of media operations, including the establishment of reform concepts and the implementation of mechanism innovation and drive.

In conclusion, the existing literature largely focuses on the operational mechanisms among businesses, without regarding the express delivery network. Consequently, a unified evaluation system has not yet been established. This paper thus proposes a definition of the relevant concepts of the express delivery network operational mechanism and seeks the most suitable evaluation method of the operational efficiency, constructing an index system for the evaluation of the express delivery network operational efficiency and analyzing the operational efficiency through empirical studies.

1.2.3 Reform of Express Delivery Model

Harvard professor Kuhn first mentioned the concept of "model change" in his book, *Education Model Change*, published in 1962. At that time, "model change" was seen as an innovative, groundbreaking, and subversive notion. Hedlund (1994) subsequently developed a knowledge management model and designed a heterogeneous structure comprising two management models.

The term "management model change" has been discussed in various fields. Gu (2018) noted that, in the realm of scientific and technological research, the shift in management models presents a novel challenge in both library work and information and knowledge services. Yang et al. (2015) analyzed the logical framework of educational management in four dimensions, discussing the essential characteristics of big data and its profound effect on education. He concluded that Chinese education necessitates model reform in the era of big data. Tian (2016) conducted an in-depth exploration of the characteristics and trends of China's agricultural transformation and the necessity of reform of the Chinese agricultural economic management model. Sun and Rong (2018) argued that, with the advent of the e-commerce era disrupting traditional production, supply and marketing models, the sales industry should seize the business opportunities presented by e-commerce to alter the model. Overall, many scholars, both within and outside of China, have undertaken research on the reform of the express delivery enterprise model.

In countries outside of China, scholars have achieved remarkable results in research on business model reform. Gamboa et al. (1995), Tsoukas and Chia (2002), and Osterwalder et al. (2005) employed interactive design to analyze the newspaper industry's business model reform. Kleppe (2017) explored the successful reform of large-scale enterprise management models.

In China, early research on management model reform was largely focused on the change process and its implications for enterprises. Zhang and Lin (1998) proposed the major theories of organizational reform and innovation, which included system theory, contingency theory, and behavior theory. Li (2001) argued that managers should be capable of modifying the organizational structure or work system in order to remain up-to-date with the most recent advances in enterprise management. Another scholar, Tu (2004), believed that reform was a barrier that enterprises must address in the long term. Wang (2001) discussed enterprise credit and management reform, while Ye and Qiu (2007) began to focus on the enterprise reform of their management model. They suggested that it is beneficial to construct a management model based on reality in order to achieve successful reform of enterprises. Lu and Yuan (2008) investigated the effect of the environment on enterprise organizational reform. They argued that when the environment surpasses the organization's capacity for adaptation, organizations will be compelled to reform.

As people gain a deeper understanding of the management model and management scholars continue to conduct research into model innovation in various industries, the current stage of research on model reform encompasses different aspects. Yang (2017) conducted research into the reform of the technology management model, whilst Zhang and Jiang (2017), Chen et al. (2017) and Yang (2018) researched the reform of the education model.

Internationally, a scarcity of literature exists that directly investigates the pertinent paradigm shift in terms of management innovation, as informed by the service system dynamics of the entire express delivery industry. Most of them take the form of empirical analysis and case studies, which are usually aimed at providing solutions to specific problems.

In China, research on the reform of the express delivery network model has mainly focused on the reform of the express delivery service model and the profit model of the express delivery industry. Zhao and Song (2016) noted that marketization of price formation in China is relatively low when it comes to the reform of the pricing mechanism of express delivery services and there is a presence of vicious price competition and price wars that have impeded the development of services, thereby necessitating an urgent reform of the pricing mechanism. Li et al. (2013) proposed a step-by-step pricing plan in order to address the flaws of the existing profit model in the express delivery industry, which has led to intense competition. The theoretical basis of the research draws from a range of disciplines, including humanities and social sciences, market economics, marketing, and other related fields. The authors employed various methods for the research, such as literature review, case study, topic divergence, case analysis, and interdisciplinary approaches.

1.2.4 Review on the Innovation of Express Delivery Networks Mode

Regarding the innovation of the express delivery management model, most scholars proposed that this innovation should be combined with "Internet+" technology, given that the computer network has become indispensable. Guo Qingui (2016) suggested that in the future, all physical objects will be connected to the Internet, and things will be interconnected with both things and people. Enterprises need to integrate with the Internet in order to progress. Wang Yan (2016) examined the impact of the Internet on enterprises in the new era and concluded that enterprises need to continuously reform and innovate in order to remain relevant in current times. Cheng Hua (2002) concluded that enterprises should utilize the Internet to optimize the use of resources, leading to a deeper integration. Kong Fanzhu (2017) proposed the concept of "Internet+", which can streamline the traditional business model, facilitate the transformation and innovation of traditional industries to advance industrial upgrading, and motivate enterprises to pursue a new type of development. The "Internet+" concept is built on existing Internet technology, which expands the range of linking people and objects, and connects more traditionally offline enterprises. Consequently, the success of combining "Internet+" and the express delivery industry would be conducive to the reform and pre-innovation of the network management model for express delivery companies.

Chen Qing (2014) conducted a study in China on the development model of the traditional private express delivery industry in relation to the integration of "Internet +" express delivery and provided a case analysis to offer strategies for the development of the local private express delivery industry across China from three perspectives. Ji Chenxi et al. (2015) compared the cost-effectiveness of three operating models: the centralized collection and replacement service model, the third-party terminal node alliance model, and the intelligent extraction cabinet, and identified existing issues. Zhao et al. (2016) conducted an analysis of the characteristics of the "O2O + crowd-sourcing" model in express delivery operations and identified inadequate supervision, unguaranteed service quality, and security loopholes as existing issues. Relevant suggestions and countermeasures were proposed in response. Yao et al. (2016) analyzed the potential of O2O national express delivery based on three O2O business models: "O2O + crowd sourcing service", "ON MY WAY" (Amazon), "JD Daojia", and "JD crowd sourcing". Zhang et al. (2016) investigated the feasibility of a new campus express delivery service model based on mobile phone software access to the Internet. From the perspective of shared economy and logistics, Li Haidong et al. (2017) discussed the efficient transmission and sharing of express delivery information based on shared data platforms used by teachers and students and proposed a campus express delivery service model with characteristics of shared logistics. Wang Jin (2018) conducted an analysis and summary of the development of express delivery service models, from traditional to the latest ones, such as express cabinets which have emerged in recent years. On this basis, he compared and analyzed existing terminal distribution service models and identified the impacting factors.

Lee et al. (2008) proposed an applicable method to design a rapid express delivery service network to assist service centers in integrating terminals or extending deadlines, thereby maximizing incremental profit for couriers. Zhuang et al. (2016) proposed the Shapley value method and an improved profit distribution model and discussed the problem of profit distribution after the establishment of a university express delivery alliance. Cui Xiaoyun (2016) conducted an analysis of the impact of the Internet of Things on the profit model of logistics enterprises in China, based on an examination of the logistics infrastructure, organization and information networks. Chen Zhixin and Du Xinjun (2016) further explored this, noting the shortcomings of the traditional express industry profit model and proposing that, in the context of the "Internet+" trend, express delivery companies should innovatively transform, utilizing new technologies such as cloud computing, interconnection, big data and artificial intelligence, in order to create a new profit model for industrial integration and expansion.

In conclusion, scholars both in and outside of China, when studying the operation or management mode of express delivery enterprises in the era of the Internet, tend to analyze the current state of the industry, predict future development trends, and identify the issues encountered by express delivery enterprises. Subsequently, they offer corresponding solutions. Despite the fact that there have been some investigations into the existing problems in express delivery network management, there has been a lack of in-depth research and analysis. Additionally, there is a dearth of comprehensive reviews of the management modes of express delivery enterprises in China. Furthermore, there has been a lack of exploration into contemporary management modes which may be suitable for the mode change and innovation of express delivery enterprises in China. Concerning the reform of the express delivery network management mode, there are relatively few relevant studies that have taken the form of monographs, and even fewer have focused on the corresponding targeted mode changes and mode innovations of the express delivery network.

1.2.5 Evaluation of the Developing Level of Regional Express Delivery Networks

After years of rapid development, the express delivery industry has seen a marked increase in both its scale and level. This industry has become deeply intertwined with agriculture, manufacturing, and e-commerce, and has become an integral part of people's daily lives. Although the development of the express delivery industry is increasing, its regional distribution is unbalanced. With a decrease in the number of express delivery enterprises in the eastern, central, and western regions of China, the structure of the industry needs to be improved. Consequently, scholars have taken note of the problem of unbalanced regional development and have conducted empirical analysis of the existing problems in China from multiple perspectives. Scholars have conducted both theoretical and practical research into how the level of informatization development affects the express delivery industry, how the express delivery industry and e-commerce coordinate with each other, and the regional imbalance of the express delivery industry.

Scholars have used various analytical methods to conduct research from different angles targeting the regional imbalance problem occurring in the development of China's express delivery industry. Xie Wentao (2009) conducted an empirical study on the level of logistics development in China's provinces and cities using factor analysis, which revealed the differences in the logistics development in each province and city across China. Furthermore, Xie Wentao conducted a systematic analysis of the main factors influencing the imbalance of development in each region. Li and Liu (2011) conducted research on the growth of the express delivery industry in western China and studied the causes of the imbalance, such as the level of urbanization, new consumption habits, and the availability of express services, among others. Duan (2015) developed an evaluation system of factors impacting the express delivery industry from the perspectives of economic growth, informatization, infrastructure, and human resource utilization. Employing grey correlation analysis methods, he quantitatively examined the primary factors influencing the development of China's express delivery industry. He concluded that economic level is the determining factor for the progress of the express delivery industry. Wang (2016) grouped regions in China in order to measure the imbalance in the development of the express delivery industry, and concluded that the inter-provincial imbalance in the industry across China is relatively severe and deteriorating. Qin et al. (2015) applied the DEA model to evaluate and optimize the comprehensive efficiency of the express delivery industry in Guangxi Province.

Secondly, scholars have conducted empirical analyses from different perspectives concerning the mutual influence between the express delivery industry and e-commerce. For example, Atif (2004) studied how the express delivery industry could affect the rapid development of e-commerce in terms of express delivery prices, service quality, and distribution efficiency. Zhou (2012) discussed the prominent problems of China's express delivery industry in the context of the rapid growth of e-commerce and proposed six measures for the development of express delivery enterprises across the country. Yu and Du (2013) employed the SERVQUAL model to conduct a fuzzy evaluation of the express delivery service industry in the online shopping environment. By combining qualitative and quantitative methods, they found that the express delivery service quality score in the online shopping environment at the emerging stage was low. Zhang and Zhu (2015) applied a Grey Correlation Analysis to examine the related indices of the development of China's express delivery industry from 1991 to 2013. They concluded that the online shopping indicators of Internet development had the highest correlation to the express delivery industry. Chen Bin (2016) employed a Vector Autoregressive (VAR) model in an empirical study and determined that there is a mutually beneficial relationship between China's express delivery industry and e-commerce, indicating a trend of coordinated development. However, the lagged development of the industry has a negative impact on further coordination between the two.

In addition, with the continuous enhancement of China's comprehensive informatization, the development of the express delivery industry has been greatly improved across the country. Scholars have studied the effect of informatization on China's express delivery industry, using varying analysis methods. For example, Bu Xinhua et al. (2012) posited that the gradual enhancement of informatization is an important factor in improving the rapid development of the express delivery industry. Yuan Chunhui et al. (2016) constructed an informatization standard system of the express delivery industry in China, and proposed suggestions from five aspects for its implementation.

In conclusion, experts and scholars have undertaken a systematic empirical analysis on the growth of China's express delivery sector using national and regional data in China. As for the current research status, there is more qualitative analysis and less quantitative analysis. Consequently, further quantitative analysis on the development of the regional express delivery industry is needed. The authors of the book construct an evaluation index system for the urban express delivery industry development by carrying out a qualitative analysis of the industry with empirical evidence and reliable data. This evaluation system is applied to comprehensively evaluate and analyze data of 31 provinces, municipalities, and autonomous regions in China. Thereby, the important impact factors on the development of the express delivery industry in China are examined and an important theoretical basis for improving the industry is established.

1.3 Contents and Innovation Points

1.3.1 Research Contents

The book is divided into eleven chapters, with the first chapter serving as an introduction and the second to tenth chapters comprising the main body of empirical studies. The eleventh chapter is the conclusion.

The second chapter outlines the pertinent concepts and theories surrounding express delivery network management, including the concept of express delivery, its primary characteristics, classification, the correlation and distinction between express delivery and logistics, and the research scope of express delivery network management.

The third chapter examines the development process and current state of the express delivery industry in China. It summarizes the development status of the industry in countries other than China, compares the development paths of the industry in and outside of China, and draws lessons from it. The authors analyze the current state of the Chinese express delivery market and enterprises from the perspective of market structure and scale, service capabilities, service quality, industrial synergy level, and competitive environment, while major express delivery enterprises, their development levels, and the construction of express delivery networks are discussed.

Chapter 4 investigates the characteristics and network structure of express delivery traffic flow. Building on previous research, the book summarizes the topological structure mode of express delivery networks and its major manifestations. Additionally, it examines the material composition of express delivery networks, and elucidates the regional structural properties. In terms of temporal and spatial distributions, the spatial and temporal dynamic characteristics of express delivery traffic flow are analyzed. Finally, the pattern of express delivery traffic flow is summarized.

In Chapter 5, research on the optimization of express delivery networks based on an intelligent algorithm is reported. The heuristic algorithm in express delivery network optimization is examined in detail. The advantages and disadvantages of the adaptability of numerous algorithms are explored by summarizing and analyzing the corresponding theories and modes. Following that, the ant colony algorithm and the simulated annealing algorithm are selected to optimize the simulated routes of the express delivery network in Shaanxi Province, demonstrating that the two algorithms can be effectively utilized in the actual operation of the network.

In Chapter 6, factor analysis is used to identify the key components of logistics express delivery networks. This chapter reviews the suitability of different methods by comparing commonly-used evaluation methods and factor analysis, selecting an appropriate analysis method and constructing a mathematical model for factor analysis. It also provides an overview of the factor analysis process. Specifically, the maximum variance, maximum equilibrium value, and regression evaluation methods are summarized. Upon comparison, a suitable evaluation method is chosen. Finally, through empirical analysis and the factor analysis of thirteen indicators in four aspects, the main factors that influence the development of the express delivery industry and the distribution of regional logistics are identified.

In Chapter 7, the operation mechanism of express delivery networks is systematically expounded from six aspects: power, transmission, distribution, organization management, promotion and guarantee mechanism, and the operation mode of each mechanism is sorted out. This redefines the connotation of the operation mechanism of express delivery network and establishes a theoretical framework for the operation mechanism of express delivery networks.

In Chapter 8, the authors employ a three-stage DEA method to analyze the operational efficiency of the express delivery industry. Initially, the BCC model of DEA is used to measure the operational efficiency of the express delivery industry in 31 provinces of China in 2017 based on cross-sectional data, leading to the empirical result. Subsequently, the SFA model is used to eliminate the effect of environmental factors and random errors. Finally, the adjusted input variables are investigated, and the BCC model is employed again to calculate the operational efficiency, which is then evaluated.

In Chapter 9, the current operational mode of express delivery enterprises is analysed and organised, and the relevant factors that influence the change of management mode of express delivery networks are further explored and summarised from five perspectives: market, price, logistics facilities, distribution site selection, and policy factors. Subsequently, based on these various factors, the authors suggest reforming the management mode, which includes four components: management mechanism, network system, transportation management and services. Finally, some novel management concepts are proposed.

In Chapter 10, the authors analyze the necessity of network management innovation for express enterprises from two perspectives—the macro environment and industry situation—and construct an ecological framework of management innovation for express delivery networks. A model of express delivery network management mode with the characteristics of circulation and self-renewal is proposed, which is named the Diversified Ecological Management Mode. Its main components are discussed, which include platform construction management mode of the Internet of Things + big data, network data mining + information resource optimization, network system framework + customer-focused and personalized services. Innovative strategies related to managing express delivery networks are then given on this basis.

Chapter 11 addresses the concluding remarks and research outlook of the authors. It summarizes all of the conclusions drawn in the book and offers prospective research avenues.

1.3.2 Innovation Points

(1) This book establishes an evaluation index system for the development of the express delivery industry. By analyzing various influencing factors, thirteen

indicators, which reflect the macroeconomic environment, system investment, the development of the express delivery industry, and the level of informatization, are selected. Factor analysis is used to extract key influencing factors, such as economic base factor, input factor, and industry development factor.

(2) An optimization model of express delivery networks based on an intelligent algorithm is constructed, taking into account the complexity of such networks. The ant colony algorithm and simulated annealing algorithm are applied to the optimization of express delivery networks, as a result of the analysis of emerging problems. Subsequently, the optimization model of express delivery networks is formulated. The simulation route optimization of the express delivery logistics network in the surrounding areas of Shaanxi province is conducted, and the performance of different algorithms in network optimization is compared by contrasting the optimized routes of the two algorithms. The results indicate that both algorithms can effectively address the optimization problems.

(3) A novel approach to evaluating the operational efficiency of express delivery services is proposed, based on a comprehensive three-stage Data Envelopment Analysis (DEA) method. The empirical analysis results demonstrate the accuracy and efficacy of this method, thereby contributing to the existing body of knowledge on the operational efficiency of express delivery services in China.

(4) The management mode of express delivery networks has been innovated. Through analysis, the issues encountered in the process of express delivery network management have been identified. The authors have proposed a novel management concept of express delivery networks, constructed an ecological architecture mode for the management innovation in this area, and suggested a multi-functional ecological management mode with certain circulation and self-renewal capabilities. This book probes the fundamental pathway to attaining innovation in the management mode of express delivery networks, effectively advancing the reform and long-term development in this sector.

1.4 Research Map

Firstly, the book expounds on the research background and significance by reviewing the previous research on network optimization, operational mechanism, and mode change of express delivery networks. It then identifies the concepts of express delivery, express delivery network management and its relevant ideas, defines the scope of research, and proposes network optimization theory, operation mechanism theory, and mode change theory. Subsequently, beginning with the history and current situation of the express delivery industry, the book integrates the architecture mode of express delivery network traffic flow and its characteristics and optimizes express delivery networks using the ant colony algorithm and the simulated annealing algorithm. Once the mathematical model of factor analysis is established, the key

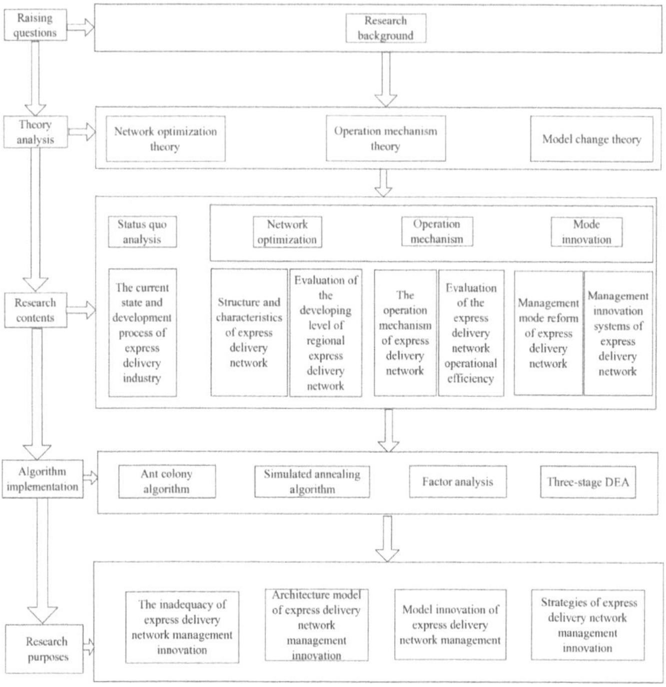

Fig. 1.1 The Research Framework (*Source* the Author)

elements of the express delivery logistics network are extracted through factor analysis. A theoretical framework of the express delivery networks' operation mechanism is constructed and the operational efficiency of the industry is analyzed using a three-stage DEA method. The authors identify issues in the existing express delivery network management models and recommend modifications. Policy suggestions for management innovation are provided. The book is concluded with a summary and outlook. The research framework of the book is illustrated in Figure 1.1.

References

Armacost, A. P., Barnhart, C., & Ware, K. A. (2002). Composite variable formulations for express shipment service network design. *Transportation Science, 36*(1), 1–20.

Armacost, A. P., Barnhart, C., Ware K. A., & Wilson, A. M. (2004). UPS optimizes its air network. *Interfaces, 34*(1), 15–25.

Atif, Y. A. (2004). A distributed and interoperable object-oriented support for safe e-commerce transactions. *Information and Software Technology*, 46(3), 149–164.

Bai, M. G., & Zhu, J. F. (2006). Comparative study on fully-connected and hub-and-spoke airline networks. *Systems Engineering Theory and Practice, 26*(9), 113–117.

Barnhart, C., & Schneur, R. R. (1996). Air network design for express shipment service. *Operations Research, 44*(6), 852–863.

Braekers, K., Caris, A., & Janssens, G. K. (2012). Time-dependent routing of drayage operations in the service area of intermodal terminals. *The 14th International Conference on Harbor, Maritime & Multimodal Logistics Modelling and Simulation.*

Brueckner, J. K., & Zhang, Y. (2001). A model of scheduling in airline networks: how a hub and-spoke system affects flight frequency, fares an welfare. *Journal of Transport Economics and Policy, 35*(2), 195–222.

Bu, X. H., Li, L., & Wang, H. Z. (2012). Study on the influence of informatization development on service level of express delivery industry. *China CIO News*, 06, 117–118.

Chen, B. (2016). Research on the interactive relationship between e-commerce and express delivery industry: dynamic empirical analysis based on VAR model. *Journal of Fujian Normal University (Philosophy and Social Science Edition)*, 1, 63–69+82.

Chen, L., Wang, H. B., Sun, H. T., & Liu, C. X. (2017). The return of MOOCs in China and the direction of reform of teaching service mode in colleges and universities. *China Educational Technology*, (8),1–6+12.

Chen, Z. X., & Du, X. J. (2016). Research on new profit model of express delivery enterprises under Internet+. *Pioneering with Science & Technology Monthly*, 29(21), 24–25.

Chen, Q. (2014). Discussion on development mode of regional private express industry: based on sample enterprises from Central-Zhejiang City Group. *Logistics Technology*, 33(1), 245–247+317.

Cheng, H. (2002). The impact of the internet on business models. *Commercial Research*, (3),140–141.

Chestler, L. (1985). Overnight air express: Spatial pattern, competition and the future in small package delivery services. *Transportation Quarterly, 39*(1), 59–71.

Cui, X. Y. (2016). Analysis of profit model of logistics enterprises based on Internet of Things. *Journal of Commercial Economics*, 28, 36–37.

Cunha, C. B., & Silva, M. R. (2007). A genetic algorithm for the problem of configuring a hub-and-spoke network for a LTL trucking company in Brazil. *European Journal of Operational Research, 179*(3), 747–758.

Dai, T., & Huo, J. Z. (2010). Research on express shipment network design based on non-hub network. *Computer Engineering Application*, 46(5), 5.

De Camargo, R. S., Miranda Jr, G., Ferreira, R. P. M., & Luna, H. P. (2009). Multiple allocation hub-and-spoke network design under hub congestion. *Computers & Operations Research, 36*(12), 3097–3106.

Dijkstra, E. (1959). A note on two problems with graphs. *Numer Math*, (1), 269–271.

Ding, J. C., Tong, W. J., & Huang, Z. X. (2008). Problems and innovation of cooperative mechanism between higher vocational college and enterprise. *Research in Educational Development*, (17), 67–70.

Duan, S. L. (2015). Empirical study on influence factors of Chinese Express industry development. *Logistics Engineering and Management*, 37(1), 171–173.

Fan, J. H., & Wang, X. F. (2006). vehicle routing optimization problem based on immune algorithm. *Computer Engineering and Applications*, (4), 210–212+217.

Feng, X. Z., & He, Z. L. (2010). Research on the operating mechanism of the organizational knowledge innovation system. *Information Studies: Theory & Application*, 33(1), 15–19.

Fisher, M. L. (1994). Optimal solution of vehicle routing problems using minimum KK-trees. *Operations Research, 42*(4), 626–642.

Floyd, R. W. (1962). Algorithm 97 shortest path. *Communications of the ACM*, *5*(6), 345.

Fu, Y. Y. (2009). *The design of supply chain networks and fuzzy multi-objective optimization based-on fuzzy theory* [Doctoral dissertation, Zhejiang University].

Gamboa, A. L. S., Filipe, E. J. M., Brogueira, P. (1995). Coordination in software development. *Communications of the ACM, 38*(3), 69–81.

Ge, X., Yu, B., & Jin, Z. H. (2013). Optimization of regional express network based on time threshold and tariff discounts. *Journal of Dalian Maritime University, 39*(2), 73–77+81.

Gu, L. P. (2018). Data management services in the transition of research model: an approach of implementing open access, open data and open science. *Journal of Library Science in China*, 44(6), 43–58.

Guo, D. H. (2016). *Internet new business model.* Beijing: China Machine Press.

He, B. Q., Li, K. P., & Cheng, X. X. (2019). The research on the optimization of courier companies "last-mile" express pickup and delivery process. *Operations Research and Management Science*, (1), 27–34.

Hedlund, G. (1994). A model of knowledge management and the n-form corporation. *Strategic Management Journal*, 15(S2), 73–90.

Horner, M. W., & O'Kelly, M. E. (2001). Embedding economies of scale concepts for hub network design. *Journal of Transport Geography, 9*(4), 255–265.

Hu, X. D., & Zhang, H. S. (2001). Applying hierarchy model to multiobjective decision making on postal transportation operations. *Systems Engineering and Electronics*, (9), 74–77.

Huang, J. H. (2009). Analysis on the complexity and robustness of express distribution network—a case study of an express enterprise. *Journal of Southwest Jiaotong University (Social Sciences), 10*(6), 98–102.

Jeong, S. J., Lee. C. G., & Bookbinder, J. H. (2007). The European freight railway system as a hub-and-spoke network. *Transportation Research Part A: Policy and Practice, 41*(6), 523–536.

Ji, C. X., Liu, B. D., & Zhuo, X. Y. (2015). Research on innovation and comparison express delivery terminal operating model on campus in the e-commerce environment. *Business*, 39, 263–264.

Jiang, J. C., Wu, L. X., Yang,Y. Z., & Li Z. F. (2014). Self-adaptive Algorithm for network maximum flow problem: SAPR Algorithm. *Application Research of Computers*, (10), 2969–2973.

Karkazis, J., & Boffey, T. B. (1995). Optimal location of routes for vehicles transporting hazardous materials. *European Journal of Operational Research, 86*(2), 201–215.

Kleppe, A. G. (2017). First European workshop on composition of model transformations—CMT 2006. *Journal of Zoological Systematics & Evolutionary Research*, 33(3–4), 42–47.

Kohl, N., & Madsen, O., (1997). An optimization algorithm for the vehicle routing problem with time windows based on Lagrangian relaxation. *Operations Research, 45*(3), 395–406.

Kong, F. Z. (2017). Research on enterprise business model innovation path under the background of internet plus—a case study of manufacturing enterprises. *China Journal of Commerce*, 20, 133–135.

Kuby, M. J., & Gray, R. G. (1993). The hub network design problem with stopovers and feeders: the case of federal express. *Transportation Research Part A: Policy and Practice, 27*(1), 1–12.

Kulak, O., Sahin, Y. & Taner, M. E. (2012). Joint order batching and picker routing in single and multiple-cross-aisle warehouses using cluster-based tabu search algorithms. *Flexible Services and Manufacturing Journal, 24*, 52–80.

Lee, H. J., Ko, H. J., & Sohn, Y. H., et al. (2008). *Profit-based network design in express courier services.* International Conference on Innovative Computing Information & Control[Conference]. IEEE, New York, United States.

Li, L., & Ding, Y. Z. (2012). Optimization of hub location and allocation for hub-spoke express delivery network. *Journal of Shanghai Maritime University, 33*(2), 33–39+75.

Li, H. (2001). Carry reform through to the end. *Enterprise Management*, 10, 1.

Li, H., Qin, Q. L., & Fang, D. L. (2014). Case study of bonded logistics system in interconnection. *Economic Review Journal*, 3, 71–74.

Li, J. (2017). Research on the operation mode of modern logistics park in Linying, Henan Province. *Journal of Henan College of Finance & Taxation*, 1, 63–66.

Li, L., Yang, H. Z. (2018). Impacts of operational mechanism of network platform on new service development performance of content provider. *Science Research Management*, 39(1), 18–25.

Li, L., Zhang, M., & Wang, J. (2013). Research on improving the price formation mechanism of express industry in China—a brief discussion on the application of step method in express pricing. *Price: Theory & Practice*, (6), 54–55.

Li, M., & Liu, Y. (2011). Comparative study of courier industry in eastern and western regions. *Value Engineering*, 30(4), 24–25.

Li, N., Zou, T., & Sun, D. B. (2004). Particle swarm optimization for vehicle routing problem, *Journal of Systems Engineering*, (6), 596–600.

Li, Z. W., & Zhang, X. M. (2006). Study on the optimization of logistics distribution network based on dynamic sweep and Ant Algorithm. *Journal of Engineering Management*, (4), 9–12.

Lin, C. C., & Lin, J. S. J. (2007). The multistage stochastic integer load planning problem. *Transportation Research Part E: Logistics and Transportation Review, 43*(2), 143–156.

Lin, F. (2017). Drive, transformation and innovation: SMG's institutional reform concept and path. *TV Research*, (7),15–17.

Lu, W. T. (2016). Regional express network optimization based on time threshold. *Journal of Wenzhou University (Natural Science)*, 37(2), 46–54.

Lu, X. W., & Yuan, Y. J. (2008). Process of industrial evolution based on enterprise organization reform. *Journal of Dalian University of Technology (Social Sciences)*, 29(4), 31–35.

Marie, S., Didier A., & Daniel R. (2012) Reducing intermodal transportation impacts on society and environment by path selection: a multiobjective shortest path approach. *IFAC Proceedings Volumes, 45*(6), 505–513.

O'Kelly, M. E. (1987). A quadratic integer program for the location of interacting hub facilities. *European Journal of Operational Research, 32*(3), 393–404.

Osterwalder, A., & Yves, P., & Chirstopher, L. T. (2005). Clarifying business models: origins, present, and future of the concept. *Communications of the Information Systems, 15*(5), 1–25.

Padberg, M., & Rinaldi, G. (1991). A Branch-and-Cut Algorithm for the resolution of Large-scale symmetric traveling salesman problems. *SIAM Review, 33*(1), 60–100.

Pels, E., Nijkamp, P., & Rietveld, P. (2000). A note on the optimality of airline networks. *Economics Letters, 69*(3), 429–434.

Qin, L. G., Wang, N. N., & Li, J. (2015). A DEA measurement and optimization of comprehensive efficiency of regional express delivery industry: in the case of Guangxi. *Logistics Technology*, 34(9), 108–111.

Ren, M. M., & Yang, C. (2007). Optimized design of waste joint recycling logistics network. *Statistics and Decision*, (14), 34–36.

Smirti, M., Boubert, A., Calloud, V., & Papson, A. (2007). Modeling the logistics of FedEx International Express. *National Urban Freight Conference* [Conference]. University of California at Berkeley Institute for Transportation Studies, Long Beach, United States. Retrieved from http://www.metrans.org/nuf/2007/documents/Smirti-000.pdf.2090928.2007.

Sun, H. Q., & Rong, N. N. (2018). Reform of retail enterprise management mode in e-commerce environment. *Journal of Commercial Economics, 4*, 100–102.

Tan, K. C., Chew, Y. H., & Lee, L. H. (2006). A hybrid multi-objective evolutionary algorithm for solving truck and trailer vehicle routing problems. *European Journal of Operational Research, 172*(3), 855–885

Tan, W. H., & Xu, C. G. (2016). Analysis of the operation mechanism of financial companies under the guidance of coordination and control. *Finance & Accounting*, (8), 50–52.

Tian, T. (2016). Reform of contemporary agricultural economy and transformation of management mode. *Agricultural Economy, 5*, 36–38.

Tsoukas, H., & Chia, R. (2002). On organizational becoming: rethinking organizational change. *Organization Science, 13*(5), 567–582.

Tu, C. (2004). *Management change—carry out management innovation to the end*. Beijing: China Logistics Publishing House.

Wang, B. Y. (2016). An analysis on regional disparities and dynamic evolution of express industry development in China. *China Business and Market, 30*(2), 36–44.

Wang, J. (2018). Classification and influencing factors of terminal delivery service mode of urban express delivery. *Journal of Commercial Economics, 8*, 81–85.

Wang, L., Chen, Y. Z., Fan, Y., & Xu, Z. W. (2016). Analysis of the operation mechanism of the national postal industry and its efficiency: based on the evaluation of the two-stage DEA. *Journal of Beijing Jiaotong University (Social Sciences Edition), 15*(3), 104–113.

Wang, W., & He, G. W. (2008). Research on the community criterion and the operational mechanism of rural saving and credit cooperatives. *Issues in Agricultural Economy, 9*, 23–28+110.

Wang, Y. (2016). *Enterprise reform in the internet era*. Beijing: Tsinghua University Press.

Wang, Z. T. (2001). *Enterprise informatization and management reform*. Beijing: China Renmin University Press.

Wasner, M., & Zaepfel, G. (2004). An integrated multi-depot hub-location vehicle routing model for network planning of parcel service. *International Journal of Production Economics, 90*(3), 403–419.

Wu, Y. T., & Lin, R. (2010). Discussion on the operation mechanism of logistics public information platform. *Fujian Transportation Science and Technology, 5*, 116–119.

Wu, Z. F., Lin, M. H., & Chen, W. (2007). Based on characteristics of regional economy of Fujian province aquatic products cold-chain logistics development research. *Journal of Harbin University of Commerce (Social Science Edition), 1*, 100–107.

Xiang, G. P., Yu, J. H., & Huang, W. (2016). The international experience and implication of the operation mechanism of technology business accelerator. *Science & Technology Progress and Policy*, 33(20), 30-36.

Xie, W. T. (2009). Regional logistics and empirical study in China's provinces and cities. *Modern Business Trade Industry, 21*(6), 69–70.

Xu, Z. B., Gu, T. L., & Zhao, L. Z. (2005). A novel symbolic ADD Algorithm for maximum flow in networks. *Journal on Communications*, (2), 1–8.

Xu, J., Huang, D. L., & Xiao, Y. (2007). Research on the construction and operation mechanism of regional logistics information network system. *China Water Transport*, 11, 126–127.

Yan, X. T., & Wu, M. Q. (2009). Research on multi-objective optimization for shortest path algorithm based on GA. *Control and Decision, 24*(7), 1104–1109.

Yang, C. P., Zheng, S. J., Dang, Y. J., & Yang, Q. (2016). Optimization of express delivery network based on delivery timeliness and connection cost. *Systems Engineering-Theory & Practice, 36*(8), 1983–1992.

Yang, C. P., Zheng, S. J., Dang, Y. J., & Cui, J. Q. (2015). Optimization of express network based on the constraint of delivery time and node flow. *Systems Engineering, 33*(11), 53–59.

Yang, H. B. (2017).The change of science and technology management mode from the perspective of innovation ecology. *Scientific Management Research, 35*(6), 43–46.

Yang, H., Nie, Y., Zhang, H., Di, Z., & Fan, Y. (2009). Insight to the express transport network. *Computer Physics Communications, 180*(9), 1511–1515.

Yang, Z. G. (2018). On the reform of the higher education governance model in France-historical institutionalism: a double examination of institutional isomorphism theory. *Contemporary Education Sciences, 2*, 75–78.

Yang, Z. Z., & Zou, W. Q. (2011). Network optimization of RPTB express based on Ant Colony Algorithm. *Journal of Transportation Systems Engineering and Information Technology, 11*(1), 90–95.

Yao, B. H., Cui, Z. W., Xu, M. Y., & Meng, L. (2016). Research and promotion of national express business model based on O_2O model. *Computer Knowledge and Technology, 10*, 285–286.

Ye, M. H., & Qiu, M. (2007). The changing management based on reality. *Business and Management Journal, 19*, 94–96.

Ye, Y. H., & Wang, L., et al. (2004). The optimal design method for the Nation's Post Network. *Journal of Industrial Engineering and Engineering Management, 2*, 39–43.

Yu, B. Q., & Du, G. W. (2013). SERVQUAL-Model-Based Fuzzy Evaluation of express service quality in online shopping environment. *Industrial Engineering Journal, 16*(2), 127–133.

Yuan, C. H., Du, H., & Zhang, Y. (2016). Informatization standard system of express industry. *Journal of Beijing University of Posts and Telecommunications (Social Sciences Edition), 18*(5), 61–66+94.

Zhang, X. F., & Lin, X. F. (1998). Reform of enterprise organization mode in the era of knowledge economy. *Journal of Xi'an Jiaotong University (Social Sciences), 4*, 36–38.
Zhang, W. B. (2015). *The air express network planning study of F Company* [Master dissertation, East China University of Science and Technology].
Zhang, C., Pan, T., & Tian, R. F. (2016). Based on WeChat public platform, a Study on the micro-express service mode of campus: a case study of Tianjin Agricultural College. *Education and Teaching Forum, 32*, 85–86.
Zhang, H. B., & Nie, Y. C. (2009). Analysis on network system of China express delivery enterprises. *Logistics Technology, 28*(9), 35–37.
Zhang, J., Wu, Y. H., Liu, P., & Wang, Y. Y. (2008). Hybrid hub-and-spoke network planning of road express freight. *Journal of Shandong University (Engineering Science), 38*(5), 6–9+13.
Zhang, L. X., Zhao, Y. Q., & Pan, F. Q. (2009). Optimal path solution of road network based on Hopfield neural network. *Journal of Harbin Institute of Technology, 41*(9), 222–224.
Zhang, L., & Zhu, H. Q. (2015). Grey correction analysis of influencing factors of China's express delivery industry. *Journal of Commercial Economics, 29*, 122–124.
Zhang, Y. X., & Jiang, H. (2017). Discussion on the reform of university teaching mode using micro-class as an opportunity. *The Party Building and Ideological Education in Schools, 6*, 35–37.
Zhao, J. Y., Wang Y., & Hu, D. W. (2004). Ant Colony Algorithm of distribution network planning, *Journal of Traffic and Transportation Engineering*, (3), 79–81.
Zhao, J., Zhang, J. J., & Yan, C. H. (2016). Research on hybrid hub-spoke express network decision with point-to-point direct shipment. *Chinese Journal of Management Science, 11*, 58–65.
Zhao, L. M. (2004). Study on the operation mechanism of digital publishing alliance: based on social network analysis. *Publishing Journal, 22*(6), 74–77.
Zhao, X. P., & Song, X. (2016). Application of O2O+crowdsourcing in logistics field. *Logistics Technology, 8*, 36–38.
Zhao, Y. Z., Zhang, H. F., & Wang, F. (2016). Research on the pricing mechanism reform of China's express service and relevant suggestion. *Prices Monthly, 4*, 12–17.
Zhou, X. L. (2012). A discussion on express industry development under the circumstances of e-commerce rapid development. *Enterprise Economy, 31*(5), 110–113.

Chapter 2
Concepts and Theories Related to Express Delivery Network Management

2.1 Concepts Related to Express Delivery Network Management

2.1.1 Express Delivery

Express delivery refers to a process in which an express delivery company collects packages from a sender, delivers them to a designated place within a specified time frame, and hands them over to a receiver (Tan, 2015). It is characterized by timeliness, service, safety, specialization, and informatization (Chen, 2015). According to Chapter IX of the Postal Law of the People's Republic of China, express delivery is defined as "delivery activities that are quickly completed within the committed time in effect". The express delivery network refers to a network established by enterprises and composed of distribution and collection centers, as well as pickup and delivery points along traffic routes, which facilitates the smooth flow of goods. This network is the material basis for the regular operation of the express delivery industry (Liu, 2016). The structure of the express delivery network, the sorting capacity of the nodes, and the planning of distribution routes all have a substantial impact on the transportation efficiency of express delivery.

Figure 2.1 illustrates the collaborative process of an express delivery network and its operations. Customers who wish to send packages can contact the couriers of a nearby express delivery outlet either by phone or by placing an order online. Additionally, customers can also send packages directly to a regional express delivery outlet. Upon the arrival of the package at the express delivery outlet, the company will package it and attach a logistics tracking number. When the number of packages to be delivered accumulates to a certain amount, they will be uniformly transported to the local city's distribution center. Packages will be sorted in the centre according to their various destinations and types, before being transported to the next station by cars, high-speed trains, airplanes, etc. Should there be no direct shipping route

P. Li and J. Wu, *Research on Innovation of Express Delivery Network Management*,
https://doi.org/10.1007/978-981-99-8982-9_2

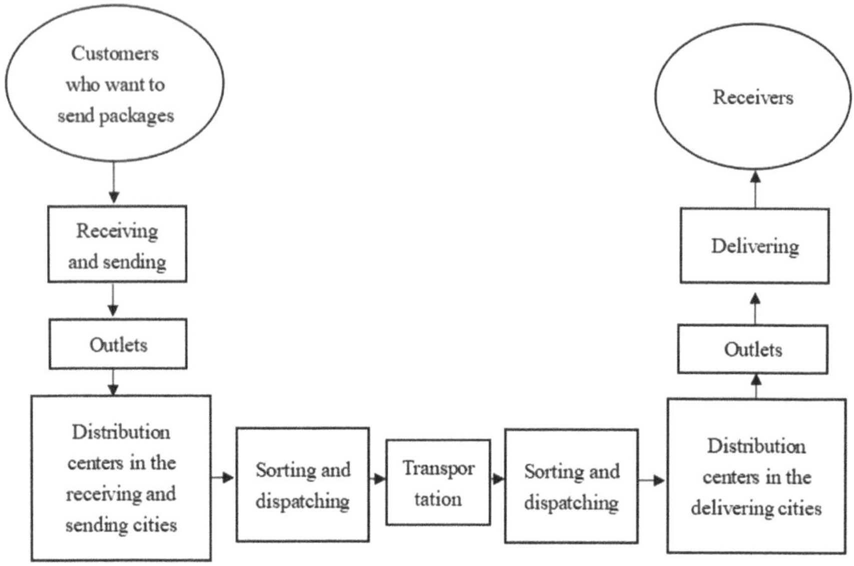

Fig. 2.1 Basic flow of express delivery in the express delivery network. *Source* The Author

between the shipping place and the receiving place, they will be sent to a transit centre for sorting and transshipment before finally arriving at the distribution centre in the receiving city. Generally, the final sorting operation will take place at the distribution centre in the destination city, and the courier will use a car, tricycle, or other means of transportation to get to the nearest express delivery outlet to reach the recipient. This process is known as 'dispatch', and customers can then collect their packages from the outlet.

2.1.1.1 Basic Characteristics of Express Delivery

The primary characteristics of express delivery include the following four points: ① Specificity of delivery; ② Uniqueness of delivery method; ③ Timeliness of delivery process; ④ Particularity of delivery organization network.

2.1.1.2 Classification of Express Delivery

Express delivery is divided into three categories, as shown in Table 2.1, based on the content of the items to be delivered, the distance of delivery, and the time limit for delivery. The content of the items to be delivered can be divided into letters, business documents, and packages (Ni, 2012). The distance of delivery can be divided into domestic and international express delivery, with domestic referring to intra-city, intra-provincial and interprovincial express delivery (Zhang, 2008). The time limit for delivery can be divided into same day delivery, overnight shipping, standard delivery, special express delivery, and general express delivery.

2.1.1.3 Differences Between Express Delivery and Logistics

According to the definition of logistics specified in China's National Standards, it can be said that logistics is the process of the flow of goods from the origin to the destination. It integrates transportation, storage, loading and unloading, handling, packaging, circulation processing, distribution, information processing, and other related functions in order to meet the actual needs.

By analyzing the definition and characteristics of express delivery, we can determine that, while both express delivery and logistics involve the physical transfer of goods through transportation, express delivery differs from logistics in terms of the objects of delivery and overall service connotations.

(1) Delivery objects: express delivery primarily handles smaller items such as business documents and packages, while logistics services cater to larger and medium-sized items.
(2) Service: express delivery typically involves either "table-to-table" or "door-to-door" pickup and delivery; logistics requires clients to bring the goods to be delivered to a logistics company, while receivers must collect the goods from the company themselves. Should clients request a door-to-door service, they will incur additional charges.
(3) Charging: express delivery charges, calculated in kilograms, are expensive yet delivery is fast; whereas logistics charges, calculated in tons, are cheap but delivery is slow.
(4) Time commitment: express delivery typically necessitates a stringent temporal commitment to accomplish the delivery mission (ordinarily within a limited duration), while logistics does not.

The differences between the two are illustrated in Table 2.2.

2.1.2 Express Delivery Networks

Express delivery network is a specialized adjunct to transportation network. It is an intricate system of numerous express delivery hub centers, all tiers of transits, express delivery outlets, transportation routes and clients, which are subject to the governance of management network and adhere to certain principles and methods. Express delivery network comprises distinct layers, generally classified as physical, layer, business layer and control layer. Physical layer enables express delivery to be expeditiously and methodically dispatched in accordance with the specifications of business layer. Control layer plays the role of monitoring, controlling, and coordinating to guarantee the normal operation of express delivery network.

Table 2.1 Classifications and characters of express delivery

Classification	Express delivery category		Characteristics
By consignment	Letters		Personal correspondence
	Business documents		Commercial contracts, engineering drawings, financial tickets, and other printed items
	Packages		Goods, gifts and other articles that can be safely posted or delivered
By the distance of delivery	Domestic express delivery	Intra-city express delivery	Being sent and received in the same city
		intra-provincial express delivery	Being sent and received within the same province
		Interprovincial express delivery	Being sent and received across the provinces
	International express delivery		Being sent and received across the countries
By the time limit for the delivery	Same day delivery		Within the specified service scope and delivery time, and promising to deliver before 8 p.m. on the same day
	Overnight shipping		Within the specified service scope and delivery time, and promising to deliver before midnight in the next day
	Standard express delivery		Within the specified service scope and delivery time, and allowing real-time tracking by custoners. Minimal delivery time: 2 days

(continued)

Table 2.1 (continued)

Classification	Express delivery category	Characteristics
	Special express delivery	Within the specified service scope and delivery time, and allowing real-time tracking by customers. Minimal delivery time: 3 days
	General express delivery	The Courier picks up the packages before the cut-off time and deliver it 2 to 6 days later

Source The Author

Table 2.2 Differences between express delivery and logistics

Content	Express delivery	Logistics
Industry involved	Postal service industry	Transportation industry
Market access	An express delivery business license must be obtained	A road transport operation permit is required
Business process	Receiving and delivering, sorting, transporting and signing for delivery	Storage, transportation
Intrinsic property	Dangerous/prohibited items are prohibited	Unconstrained
Item specification	For small packages, no more than 150 cm on either side and no more than 300 cm in length, width and height	Unconstrained
Time requirement	The sooner, the better	Unconstrained
Charge standard	Relatively high cost	Relatively low cost
Administrative department	Posts administration	Transportation Administration
Implementation standard	The national standard for Express delivery Service (GB/T 27,917.1 – 2011)	The relevant regulations of the logistics standards special planning

Source The Author

2.1.3 *Express Delivery Network Management*

The management of an express delivery network involves the planning, organizing, commanding, coordinating, and controlling of it through the application of scientific methods and specific regulations in the process of social reproduction.

2.2 Network Optimization Theory

We live in a networked society, which is a complex system composed of computer information networks, telephone communication networks, transportation service networks, material distribution networks, energy sources, etc. (Xie et al., 2009). Xie Fanrong (2009) suggested the concept of network optimization: the maximization of social and economic benefits through effective planning, management, and control of the network system.

Network optimization is a classic and essential branch of operational research, encompassing the study of material management, economic management, industrial engineering, transportation, computer science and information technology, communication and network technology, and cybernetics and military operational research.

The shortest path, minimum saturated flow, maximum flow, minimum cost maximum flow, minimum cost flow, maximum profit flow, and maximum capacity path are some of the most fundamental problems in the network optimization process. Scholars worldwide have conducted extensive research on these topics and have achieved positive results. For instance, Xu Zhoubo (2005) and Jiang Jincheng (2014), two Chinese scholars, used the ADD algorithm and the SAPR algorithm to solve the maximum flow problem, while the Dijkstra algorithm (Liu et al., 2011) and the Floyd algorithm (Zhang & Wu, 2009) are commonly employed to solve the shortest path problem.

Thomas (2006) noted that the core of the shortest path problem is, when each edge of a network has been assigned a cost based on length, cost or time, the path with the lowest total weight between the source point and the destination point is the solution to the shortest path problem. This can be applied to practical issues, such as finding the shortest transportation route, vehicle routing arrangement, factory location and layout.

In 1956, Ford and Foxon proposed the Maximum Flow Labeling Algorithm, thus establishing the "Network Flow Theory". The Maximum Flow Problem is a Combinatorial Optimization Problem with the purpose of maximizing the transportation flow in order to maximize efficiency and optimize the utilization of equipment.

Qian Songdi (1990) discussed the network's minimum-cost-with-maximum-flow problem in detail. There are paths in any network, and each one is limited by two conditions: "capacity" and "cost". Research on this problem seeks to determine how to select the path and allocate the flow along it in order to achieve the minimum cost while still allowing the maximum flow from the starting point to the end point.

Lu et al. (2003) stated that maximum profit flow is the goal of maximizing transportation profit in network optimization. A feasible profit flow can be decomposed into multiple path and circular flows, and the feasible flow's profit is equal to the sum of the profits of these path and circular flows.

In multi-objective programming models and algorithms used for studying network optimization problems, where the complexity of the problem and humans' bounded rationality are taken into account, a satisfactory or effective solution is sought when searching for the optimal solution. Zhou Shufa et al. (2004) took engineering network optimization as their research object, establishing a multi-objective planning mode that comprehensively considered the quality, schedule, cost, and resources of the project in order to achieve comprehensive optimization of the network.

2.3 Operational Mechanism Theory

2.3.1 Operational Mechanism

The operational mechanism refers to the structure, function, and interrelationship of various factors that have an effect on enterprise operations, as well as the process by which these factors act and the principles they adhere to. During the enterprise's operation, the primary mechanism guides and constrains the regulations surrounding sustaining activities that involve human resources, funds, and materials. The interaction of multiple elements in the operation process is a self-adjustment technique which leads to more efficient and ordered operational activities of the enterprise.

2.3.2 Components of Operational Mechanism

Decision-making mechanism. Enterprises should make decisions about economic activities, such as production and operation, with adequate legal person property rights. This includes establishing decision-making bodies, designating decision-making authorities, creating decision-making organizations, and establishing decision-making methods.

Incentive mechanism. The essence of the incentive mechanism is an economic benefit mechanism, which can maximize the enthusiasm, initiative, and creativity of employees.

Constraint mechanism. Enterprises must adapt to the changing environment and continuously constrain and adjust their behavior in order to survive and develop optimally. A corporate behavior constraint mechanism comprises both internal and external behavior constraints. Internal constraints include budget, audit, responsibility, and disciplinary mechanisms, while external constraints comprise those of the market, contract, law, and certain regulatory bodies.

Running mechanism or circulatory mechanism. Enterprises should ensure smooth supply, production, and sales operations by continuously adjusting their systems and improving their relevant decision-making systems. By integrating various elements into the production and operation process, enterprises can enhance their adjustment abilities in all aspects. Furthermore, by increasing their adaptability to cope with various risks and enhancing their competitiveness, enterprises can maintain their stability in the external environment and promote their development.

2.3.3 Essence of the Operational Mechanism

2.3.3.1 Objectives

The objectives of enterprise operation mechanism can be divided into three levels: the first is the operational objectives, which are determined according to the type of ownership; the second is the developmental objectives; and the third is the objectives that production workshops or employees are required to meet.

2.3.3.2 Incentive

The operation of an enterprise is analogous to that of a machine, which necessitates a specific amount of energy. Interest is the fulcrum of its expansion. Similar to the "rotation of the earth", it is propelled by the interest of "rotation".

2.3.3.3 Regulation

An enterprise's rational behaviour necessitates an excellent socio-economic environment, proper guidance, and appropriate adjustment; however, the ability to strictly regulate is more essential.

2.3.3.4 Objectivity

(1) Enterprise carries out operational mechanisms through specific management methods and business activities.
(2) The operational mechanism of an enterprise is an essential link between the enterprise and its external environment. Every enterprise has its own distinctive operational mechanism, which may vary according to the stage of the enterprise. Consequently, the operational outcomes also differ.
(3) The objective existence of an enterprise's operational mechanism must be understood, and the capacity to comprehend and implement this management mechanism will determine the success or failure of the enterprise.
(4) Internal and external factors, along with the relevant precise system surrounding an enterprise, determine whether its operational mechanism can achieve the desired effect.
(5) Enterprises operate by adhering to established operational protocols and employing precise and logical approaches.

2.3.4 Conventional and Creative Operational Mechanism

The administration's direction of conventional businesses to operate under the regulations set by the government can be thought of as an administrative unit. The conventional operational model is a subjectively-driven action of the government, not an objective requirement for productivity, which leads to difficult-to-resolve contradictions.

The creative operational mechanism is driven by interests, which is also the basis for reconstructing the operational mechanism and is in accordance with the intrinsic requirements of the market economy. The remodeling of the operational mechanism should cover the four effective mechanisms of decision-making, incentives, constraints, and operation, and should consider the operating subject, object, mode, and result.

In brief, the operational mechanism must accurately reflect its nature. It should coordinate the two primary regulatory systems of the operational mechanism and their respective sub-mechanisms in order to create a rational, comprehensive force of regulation to ensure the long-term success of an enterprise.

2.4 Model Change Theory

Thus far, mode change has not been defined precisely. Generally, it refers to events occurring within enterprises or the business realm.

Beginning with the restructuring of the retail industry, Zheng Shiming (2006) sought to understand the internal and external mechanisms of business evolution by posing the question of "what is a business model." He formulated an initial concept of "business model" and highlighted the importance of changing business model as a means of gaining a competitive edge.

Business model change involves classifying business models according to the various underlying characteristics. Integrating economic phenomena and theories, simple and comprehensible economic theories are employed to explain business model changes and to further extend the meaning of value chain theory. Enterprise transformation is an overhaul initiative that must be undertaken when growth is lagging or when problems arise or adjustments cannot be made relative to the business environment. It is essentially implementing corresponding corrections to ensure the healthy development of an enterprise, with regards to internal structure, production operation mode, and corporate culture.

Management transformation is an essential step for an enterprise to progress, and involves changing the entire management mode, approaches, and skills conducted by the enterprise. While the aim is to transform successfully, these changes may not necessarily have the desired outcome, or even fail. Transformation is a risky undertaking; however, due to the intense competition within the same industry, constant technological advancements, and the necessity for talent, it is a viable way for an enterprise to transform and stay abreast of the industry trends. Enterprise transformation and organizational change are the core components of the management reform of the express delivery network.

Organizational change is an integral part of enterprise transformation. The progression of an enterprise is an upward spiral. Integrating and adjusting resources in order to meet production and operational demands presents both challenges and opportunities to an enterprise, thus driving enterprise transformation through organizational change.

Organizational change involves the rapid adaptation, enhancement, and innovation of the components of an organization (management principles, processes, structure, staffing, corporate culture, etc.) in response to changes in both the internal and external environment of the enterprise (Li, 2019). Managing organizational change requires organizational leaders to have a firm understanding of potential obstacles and resistance. Furthermore, for each specific issue, there must be comprehensive plans and implementation steps so that problems can be successfully addressed through management reform.

2.5 Summary

First, the chapter elucidates the concept of express delivery, its categorization, and the differences between express delivery and logistics, and delineates the scope of the express delivery network. Subsequently, the authors analyze the relevant research results at home and abroad to investigate the fundamental issues in network optimization, assess the distinctions between traditional operational mechanisms and new operational mechanisms, and organize the theory of model change, which provides a theoretical basis for subsequent research.

References

Chen, L. (2015). *Information technology progress and competitive strategy choice of private express delivery enterprises*. Beijing: China Social Science Press.

Cormen, T. H. (2006). *Introduction to algorithms*. Beijing: China Machine Press.

Jiang, J. C., Wu, L. X., Yang, Y. Z., & Li, Z. F. (2014). Self-adaptive algorithm for network maximum flow problem: SAPR algorithm. *Application Research of Computers, 31*(10), 2969–2973.

Li, W. X. (2019). *A study on the reform of teaching and research group's professional autonomy* [Doctoral dissertation, East China Normal University].

Liu, J. M., Ma, S. F., & Ma, S. Q. (2011). Computation method of the dynamic shortest path based on improved-Dijkstra algorithm. *System Engineering——Theory & Practice, 31*(6), 1153–1157.

Liu, P. (2016). *Research on county location model of rural express delivery business outlets* [Master dissertation, Chang'an University].

Lu, H. S., Gao, X. D., & Wu, S. (2003). An algorithm for maximum profit flows. *Mathematics in Practice and Theory, 33*(5), 43-48.

Ni, L. L. (2012). *Study on express operation network optimization design and competitive network equilibrium* [Doctoral dissertation, Central South University].

Qian, S. D. (1990). *Operations research.* Beijing: Higher Education Press. SF Express website (2018, April 5). Retrieved from http://www.sf-express.com/cn/sc/express/express_servicemainland_area/.

Tan, X. (2015). *Analyzing and modeling express shipping service network based on logistics data* [Master dissertation, Zhejiang University].

Xie, F. R. (2009). *Research on efficient & robust algorithms for some problems in network optimization and its application in management* [Doctoral dissertation, Nanchang University].

Xie, J. X., Xing, W. X., & Wang, Z. B. (2009). *Network optimization.* Beijing: Tsinghua University Press.

Xu, Z. B., Gu, T. L., & Zhao, L. Z. (2005). A novel symbolic ADD algorithm for maximum flow in networks. *Journal on Communications, 26*(2), 1–8.

Zhang, D. Q., & Wu, G. L. (2009). Optimization of floyd algorithmic for shortest path problem. *Journal of Xuchang University, 28*(2), 10–13.

Zhang, L. (2008). *Research on net point layout of express enterprise* [Master dissertation, Central South University].

Zheng, S. M. (2006). *Business model reform.* Guangdong: Guangdong Economy Publishing House.

Zhou, S. F., & Liu, L. (2004). The multi goal optimization problem in network project. *Journal of East China Jiaotong University, 21*(2), 10–13.

Chapter 3
Development and Status Quo of the Express Delivery Industry

The express delivery industry was developed when social productivity reached a certain stage. In ancient times, there were messengers who were responsible for delivering documents or information to the emperor or king for political purposes, and also couriers serving ordinary people. In a sense, modern express delivery is a form of economic exchange with economic and social benefits. Internationally, the express delivery industry started early in developed countries in Europe and North America, and a relatively mature and complete operating model has been established. Although the industry emerged later in Japan, a unique system of development has formed thanks to the sound economic foundation in the country. This chapter will analyze and review the development of express delivery industry in the United States, Germany, and Japan.

3.1 Development of the Express Delivery Industry

The express delivery industry originated in the United States in the 1960s, and United Parcel Service (later renamed UPS) is the world's first express delivery company. The express delivery industry grew out of the postal services. In the late 1970s, the express service industry was brought into China. In 1985, China Post established EMS, an express service company, marking the start of express delivery business in China, and it developed rapidly with the help of the express delivery network and infrastructure of the postal service. China's private express delivery enterprises began in the early 1990s. In this period, the market economic system was gradually established, and the number of enterprises grew rapidly. The market economy was active then, and transactions between enterprises were increasing. At the same time, thanks to the continuous improvement of the domestic transportation network, the

P. Li and J. Wu, *Research on Innovation of Express Delivery Network Management*,
https://doi.org/10.1007/978-981-99-8982-9_3

rapid development of the means of transportation and telecommunications, especially the flexibility of the regulation on delivery service, private express delivery enterprises have been able to flourish.

3.1.1 Development of the Global Express Delivery Industry

3.1.1.1 Development of the Express Delivery Industry in the United States

At present, The United States is the leading country in the international express delivery industry. The global express delivery giants UPS and FedEx have strong competitiveness around the world. The development of the industry in the United States can be divided into four stages: the initial development; the rising of leading enterprises; the development of railway express; the development of international business.

(1) The Initial Development

A developed financial service network has not been formulated in the United States in this stage. The primary operation model is to delivery along the the fixed routes or add the categories for delivery along the fixed routes so that more clients were attracted. In 1848, the U.S. government subsidized the postal service to implement the adjustment and formulation of a unified tariff standard.

(2) The Rising of Leading Enterprises

Around 1890, the US express delivery industry entered the rising stage of leading enterprises. A large number of enterprises poured into the industry, and several large enterprises carved up the country's express delivery market, which constituted several regional operation networks. At this stage, the competition in the express delivery industry is fierce. Large-scale express delivery enterprises have established strategic partnerships to avoid fierce competition in the industry. This stage has injected new vitality into the development of the US express delivery industry.

(3) The Development of Railway Express

The railway network in the United States has improved gradually, and the express delivery industry has entered the stage of standardized operation by railway express. In 1914, the Interstate Commerce Commission (ICC) regulated the attach of express and mail vehicles to passenger trains. In 1929, the development of American railway transportation reached its golden age. Several major railway companies took advantage of their capacity to set up Railway Express Agency (REA) to operate express services. REA relied on the benefits of its rail transportation network to monopolize the US express delivery market. At the same time, through cooperation with telegraph operators, the establishment of a low-cost network in the United States has been achieved. Areas with low express delivery businesses have adopted the "one-person-department" business model to run cargo delivery by multimodal transport. Express delivery operation specifications and systems have been increasingly improved. In the late of this stage, with the development of other modes of transportation, the traffic volume of railway transportation was dispersed, resulting in a sharp reduction in railway capacity, and the market structure gradually shifted to a competitive market.

(4) The Development of the International Express Delivery Business

When the express delivery market in the United States has been developed into a stable stage, UPS (United Parcel Service) and FedEx (Federal Express) began to step into the international market. They soon became the four largest international express delivery companies along with DHL and TNT. The US express delivery industry has entered the era of global express. In this stage, express delivery enterprises jointly established a multi-network transportation layout, and the means of transportation increasingly diversified. Among the various transportation methods, air freight has become the primary one. The multi-network transportation layout has generated a variety of transportation methods, and multiple transportation methods jointly participate in the transportation of goods, which makes the express transportation process more complicated.

The co-construction of multi-network by express delivery enterprises takes a lot of time, human resources, and material on information, transport and operation networks. And the continuous update of new technologies has gradually expanded the scope of their application. However, in the development of the international express, there are still a series of problems occurring in customs, air rights negotiation, and so on. Therefore, it also needs strong inter-governmental support.

3.1.1.2 Development of the Express Delivery Industry in Japan

The rise of the express delivery industry in Japan stems from the growing demand for small package delivery by Japanese households. Transportation enterprises have transformed from traditional distribution to "home delivery" mode which developed rapidly. Express delivery in Japan takes home-to-home delivery as its main business. It operates the logistics distribution to consumers around cities, forming a mode of "home delivery express" with unique Japanese characteristics. Although the express delivery industry in Japan started late, it has developed rapidly with the continuous enhancement of its comprehensive competence. Especially in terms of automation and informatization it is one of the advanced in the world. The indispensable external conditions and internal driving forces for the development of the express delivery industry in Japan are traffic network, transportation and perfect service system. In Japan, based on traditional logistics, the development of C2C express service has urged the express delivery industry develop rapidly, and it has also promoted the development of B2B, B2C, and C2B, which has continuously expanded the scope of the industry.

Before the emergence of the express delivery industry in Japan, the delivery of small packages was mainly completed by the state-owned post and telecommunications offices and the state-owned railway enterprises. However, the delivery speed was slow and the business efficiency was low. Until 1976, the traditional cargo transportation enterprise in Japan, Yamato Transportation, identified the disadvantages of the post and telecommunications offices and the state-owned railway industry. It discovered a gap in the competition in cargo distribution and business services. On January 20 of the same year, Yamato Transportation applied for a C2C express business license to the Ministry of Transport and named the express delivery "Black Cat House Urgent". Its business focused on delivering small packages under 30 kg for individuals and families, and promised "overnight delivery" to any location. In order to increase the aggregation of goods and the traffic volume of express delivery, Yamato Transportation firstly took the Kanto region as the scope of express delivery services, continuously expanding the distribution area and adding business points. Thus a skeleton structure for goods collection and delivery was built. At the same time, the enterprise took advantage of places frequently visited by consumers, such as rice shops, hotels, and convenience stores, as service outlets where goods are collected.

The establishment of "Black Cat House Urgent" opened the history of the express delivery industry in Japan. After Yamato Transportation participated in the small cargo market in the form of express delivery, as shown in Table 3.1, other transportation enterprises also realized the potential business for small cargo. They also competed to join the express delivery market represented by small cargo, and the competition among various transportation enterprises became increasingly fierce. In particular, Sagawa Express, which participated in the express delivery business in November 1998, achieved a significant growth rate in its business. It overtook Nippon Express—the Second largest in the business—in the first half of 1999, and by 2005 there were still 34 transportation enterprises participating in the market competition. In addition, in order to strengthen market competitiveness, express delivery enterprises seeked to develop in various orientations and expanded the scope of business services. They scrambled to provide whatever services to meet customers' needs. For example, in 1983, Yomato Transportation offered to transport irregular goods such as ski equipment for consumers, and in the following year it extended the services to golf equipment. Other express delivery enterprises also developed various delivery services to meet the special needs of customers based on their merits.

Major express delivery enterprises have launched door-to-door services as well as diversified services according to customers' needs. The express delivery industry in Japan developed rapidly, with major express delivery enterprises competing with the small package delivery business of Japan Post. In 1984, there was a milestone for the express delivery market in Japan. The number of home delivery businesses provided by major express delivery enterprises exceeded the small packages business offered by Japan Post and telecommunications office, and the trend continued to rise. The traffic volume of small packages in the State-owned railway industry also decreased sharply due to the continuous increase in the traffic volume of home delivery while its operation showed a deficit. In 1986, during the privatization of

Table 3.1 The time of major Japanese companies are engaged in express delivery

Company name	The beginning time of express	Express name
Yamato Transport Co., Ltd	January 20, 1976	Yamato Transport
NIPPON EXPRESS	April 1977	Nippon Express
Seino Transportation Co., Ltd	November 1977	Seino Transport
Fukuyama Transport Co., Ltd	June 1981	Fukutsu Tracking
Sagawa Express Co., Ltd	November 1998	Sagawa Express

Source https://kuronekoyamato.co.jp/ytc/corporate/history.html
https://www.nittsu.co.jp/sbout/
http://www.seino.co.jp/
http://www.fukutsu.co.jp/
https://www.sagawa-exp.co.jp/company/

Railways in Japan, the railway's small package delivery business ended. In addition, with the rapid development of the network economy, there appeared the door-to-door delivery service at any time in June 1998, which brought convenience to customers as well as saved costs. Many express delivery enterprises survive under the challenge of meeting the specific needs of consumers. At the same time, they are also promoting the development of the express delivery industry in the face of opportunities. And since 1998, the emergence of online stores has prompted a rapid increase in the traffic volume of B2C express delivery and pushed the express delivery industry to a new level.

The nodes of express delivery network in Japan can be categorized into three parts. They are pick-up outlets, business offices, and distribution centers. As shown in Fig. 3.1, express delivery enterprises will select places frequently visited by consumers, such as rice shops, hotels, and convenience stores, as outlets for the dispatch and delivery of goods, which can provide customers with delivery and pick-up services. This mode not only improves the convenience of services but also shortens the distance between network nodes. Business offices, pick-up outlets, and distribution centers can be connected to form regional transfer stations of express delivery. Distribution centers, which sort and deliver goods to various regions, is the core of an express delivery network. An automotive transport system is composed of three parts—collection of express shipments, trunk transportation, and distribution of express shipments. Among them, trunk transportation only involves two distribution centers to send or and receive packages. This mode effectively reduces loading and unloading time. Trunk transportation is often carried on overnight to avoid traffic congestion and achieves the service goal of next-day delivery. An express delivery network with dense outlets and fast and convenient transportation mode comprehensively improve the service quality of express.

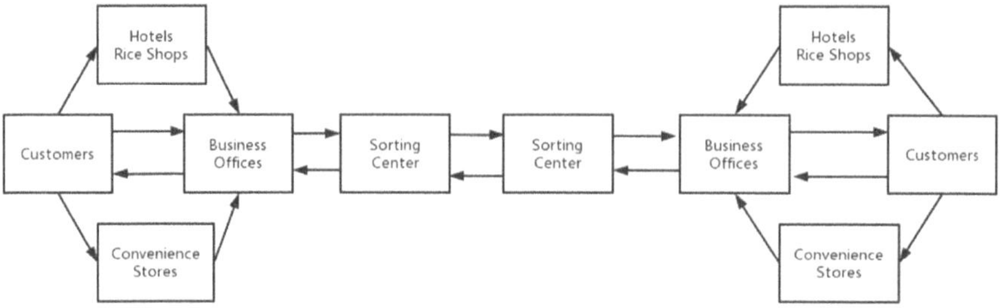

Fig. 3.1 Japan express delivery mode. *Source* The Author

3.1.1.3 Development of the Express Delivery Industry in Europe

The EU postal market opened in 1997, and two postal reform decrees were passed in June 2002. The postal reform act clearly requires: (i) the scope of post franchise must be gradually narrowed down; (ii) the universal service should continue to be maintained based on the open market; (iii) the regulatory environment should be adjusted in a timely manner according to the market situation; and (iv) a universal service permit system should be established for providers. The logistics industry is sophisticated in Germany, and the scale of it has long been ranked first in Europe. Express delivery is the focus of its development.

Germany has the most intensive transportation network in Europe. In addition to regular express services, many express delivery companies also provide multiple logistics services, such as inventory and transportation and attach importance to value-added services as well. The German express delivery industry has been a complex one which integrated e-commerce, finance, and other services. It is of the characteristics of intensification, standardization, informatization, multi-functionalization. It is no longer operated by the out-of-date single-operation mode focusing on goods transportation and postage. The primary transportation means is through the road, and the main vehicles are trucks. German Hermes provides express as well as logistics services; DHL provides basic express services for Deutsche Telekom and expands value-added services including mobile phone parts processing and other product parts processing and replacement; The primary means of transportation used by Hermes is container trucks that are attached to bridge networks. As environmental protection is greatly appreciated in Germany, transportation via bicycles is adopted for some special services which include courier transmission between cities (Lu, 2010).

To promote the rapid and healthy development of the express delivery industry, the German express sector adopts an economic system called "social market economy" which combines government macro-control and free competition, aiming at achieving a green, convenient and high-efficiency comprehensive transportation network. Under the guidance of environmental protection and coordinated development, this system takes macro-control over the construction of transportation infrastructure such as roads and railways.

On March 15, 2018, DHL said it would launch delivery services within 24 h or 48 h for leading online sellers in Chicago, New York, and Los Angeles. The service would expand to Dallas, Atlanta, San Francisco, and Washington by the end of the year, with more cities included rapidly in the coming years.

3.1.2 Development of the Express Delivery Industry in China

Since the ancient time, the national express delivery industry in China has undergone milestone events such as the issuance of the first set of stamps, the signing of the first express agency agreement, the establishment of the first professional express delivery company, and the emergence of domestic and international express delivery companies. The volume of express shipment amounts to hundreds of millions. By analyzing the development of the national express delivery industry in China, we can conclude that the development of nationwide express delivery has gone through four stages: first, the embryonic stage (1300 BC to 1948); second, the initial stage (1949 to 1985); third, the growth stage (1986 to 2005); fourth, the listing stage (2006 to present).

3.1.2.1 The Embryonic Stage (1300 B.C.–1948)

According to historical records, there were various forms of organized communication activities since the first day human beings emerge in this world. Looking into relevant materials, we learned that oracle bone inscriptions recorded in the Pan Geng period of the Yin and Shang Dynasty had described organized communication activities.

With the establishment of the Qin Dynasty, a revolutionary change in communication took place, in which the postal booth became the main communication agency (Guo, 2013). In the heyday of the Tang Dynasty, 1639 postal stations were set up with more than 20,000 people working as postal staff.

The method used in the Tang Dynasty was continued in the Song Dynasty. In addition, express delivery office was created (named *Jidipu*, also known as *Jijiaodi*, a postal station for the transmission of official documents), which was specialized in handling emergency military mails. The golden plaque in the Song Dynasty was a communication credit, which indicated that mails with it had to be delivered faster than those of ordinary express delivery The communication delivery organization was divided into three types: foot delivery, horse delivery, and express delivery.

In the Ming Dynasty, the Minister of General Affairs was the postal agency integrated with the station, transportation, and offices. The express delivery network was connected with the Shuima Station. It took counties as centers scattering around and adjoined one and another to shape the delivery road network all over the country.

In the Qing Dynasty, communication was mainly accomplished by couriers through waterways or land roads network and the express offices. The function of post offices changed dramatically due to the appearance of super express delivery by horses (called *Ma Shang Fei Di*). In March 1878, they began to receive and send letters for ordinary people and then carried out the sea and land transportation of letters and packages. This was a milestone in the postal history in China. In July of that year, the first set of postal stamps names Dalong Stamp was issued.

In 1896, Emperor Guangxu approved the establishment of a national post service, and then the Great Qing Imperial Post Office was officially established in Beijing. By 1904, the postal branches across the country had initially formed a national postal network, ushering in the era of modern postal service in China.

In 1912, the Republic of China was established, and the Great Qing Imperial Post Office was renamed the Republic of China Post.

In March 1932, the Postal Administration (Red Post) of the Chinese Soviet Republic was established. Red Post set up regulations on postage, according to which letters between the Red Army soldiers and their families were free of charge while a stamp is required for other private letters. Red Post was responsible for collecting and sending on behalf of the Postal Administration of the Chinese Soviet Republic.

On January 1, 1935, the Civil Information Bureau, formed during the Yongle period of the Ming Dynasty all suspended. In 1936, Chinese Soviet Northwest General Post Office was established through the consultation of the Central Government of Chinese Soviet and the Northwest Ministry of Internal Affairs.

3.1.2.2 The Initial Stage (1949–1985)

Upon the founding of the People's Republic of China, the Ministry of Posts and Telecommunications was established on November 1st, 1949, which was responsible for uniformly managing the postal and telecommunications affairs across the country. From December 10th to 28th of the same year, the first National Postal Conference was held in Beijing. The Postal Service of the People's Republic of China was established as a state-owned economic organization named "People's Postal Service of China". The Ministry of Posts and Telecommunications reiterated the commitment to service in 1952, and codified the principles of "fast, accurate, secure and convenient" on November 19.

Since the reform and opening up in 1978, the vitality of China's economy has been continuously stimulated and gradually integrated into the world market. For better integration with the world market, China needed to develop an export-oriented economy. But a delivery service provider that could meet the needs of foreign trade did not exist at that time. Besides, foreign companies operating in China also urgently needed convenient and efficient communication channels to connect with the world. Against this backdrop, the national express delivery business came into being. The development of the express delivery industry in China is inevitable, a reflection of the policies and the environment in the country in the early stage of reform and opening up.

China Post has always been guided by the idea of healthy and rapid development. With the deepening of reform and opening up, it has gradually improved its comprehensive strength and formed a nationwide postal network. China Post opened the international express service in 1980, domestic express mail service in 1984, and the China Express Service company (EMS) was established in 1985. In the first half of the 1990s, EMS was practically the only express delivery business operator in China. The difference between EMS and international express delivery companies is that it also undertakes domestic express delivery while engaging in international express business. EMS plays an important role in domestic and foreign markets. It has maintained a market share of more than 50% in China's international express delivery market for an extended period.

3.1.2.3 The Growth Stage (1986–2005)

In 1986, China promulgated the first Postal Law since the founding of New China, which defined the scope of the state postal monopoly. The Postal Law stipulates that postal enterprises are the sole provider of mailing and delivery services for items that can be classified as letters, and China Post is the sole entity authorized to set postal rates. This has provided China Post with the most favorable macro-policy environment. The implementation of the Postal Law has also ensured legislative support for the international development of China's express delivery industry.

With the continuous advancement of China's reform and opening-up process, China's economy entered a period of rapid development especially after the year of 1992 when Comrade Deng Xiaoping delivered a speech on his southern tour. China's Pearl River Delta and Yangtze River Delta regions have become the most active regions for China's export-oriented economy with their unique advantages, plausible mechanisms, and ideal manufacturing locations, with the continuous advancement of reform and opening up, the private economy in these regions has grown greater and stronger. As the international division of labor has grown to a higher level, today's enterprises have a higher demand for the timeliness, security, and convenience to exchange commercial documents to cope with the fierce competition in the domestic and foreign markets. Compared with other domestic express delivery companies, the price of EMS is relatively high. And with its limited network structure is, it has been operating in a less flexible pattern, which makes it unable to meet the needs of businesses for their request of delivering commercial documents. So EMS could not undertake all the business of express mail in China then.

For the above reasons, China's private express delivery industry came into being. Shanghai STO Express and Shenzhen SF Express were the first private express delivery companies established in Shanghai and Guangdong respectively in 1993. Other express delivery companies were subsequently established, such as TTK Express and Home Express in 1994, Yunda Express in 1999, and YTO Express in 2000. In 2005, UPS Sole proprietorship companies was registered to operate express

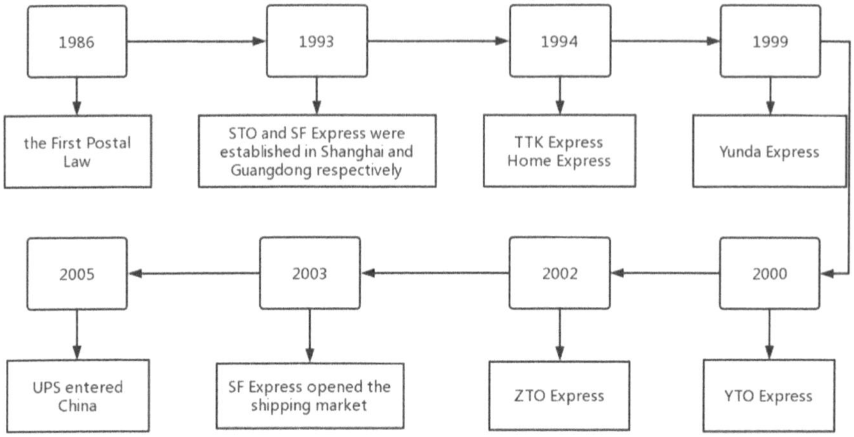

Fig. 3.2 Course of the development of the growth stage. *Source* The Author

business in China, as shown in Fig. 3.2. Since EMS could not meet the demands for high-quality express delivery services, private express delivery companies quickly filled the gap, and has secured a foothold in the express delivery market, and become an essential part of China's express delivery industry.

3.1.2.4 The Listing Stage (2006–present)

Since 2006, e-commerce has significantly promoted the development of the express delivery industry, leading to the rapid growth of private express delivery companies with an increasingly larger market share. As shown in Fig. 3.3, the national express delivery business snowballed between 2008 and 2017.

The revenue of China's express delivery industry reached 38 billion yuan in 2007, and the growth rate is higher than the average for the tertiary sector, exceeding 25%. From 2009 to 2010, Taobao began to launch Double Eleven, which made the express delivery industry developed dramatically. On December 26, 2012, high-speed rail was introduced to the express delivery industry. The arrival of the high-speed rail era reduced the cost by 50% and promoted the rapid development of the express delivery industry. Within the ten years from 2006 to 2016, the national express business volume increased from 1 billion to 31.2 billion. As it continues to grow, the full scale has jumped to the first place in the world.

In 2013, there were more than 8000 express delivery companies across the country, including more than 20 major express delivery brands. The express delivery industry has developed rapidly benefited from the growth of e-commerce. E-commerce giants have built their logistics and online express delivery platforms. The "rough era" of

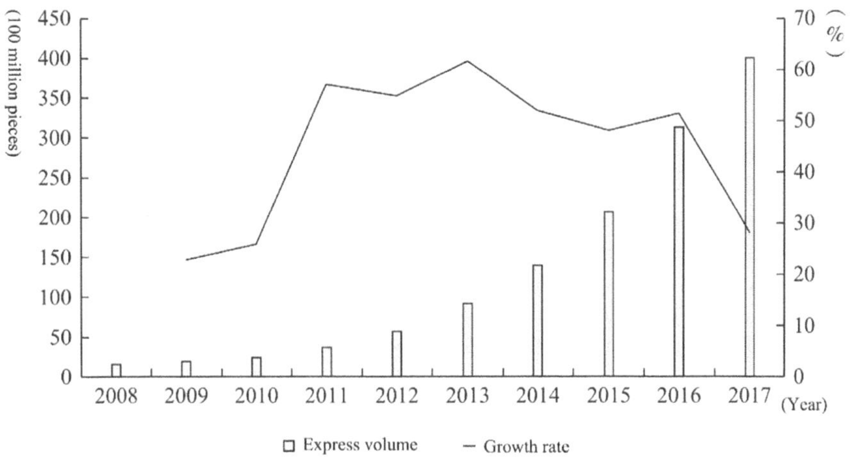

Fig. 3.3 China's express business volume from 2008 to 2017. *Source China Statistical Yearbook*

the express delivery industry has ended and it has entered into a new stage of development. "Internet +" has become the most significant opportunity for the transformation and development of the industry. From 2015 to 2016, China's express delivery industry has stepped into the capital-chasing stage.

Table 3.2 shows the listing process of some major domestic express delivery companies in China. From 2015 to 2018, major express delivery companies went public one after another. STO Express was first listed in December 2015, followed by YTO Express and ZTO Express and many others.

Table 3.2 Profiles of major express delivery companies

Company	Time	Process of listing
STO Express	December 2015	It was listed at a price of 16.9 billion yuan with the backdoor of Aidixi
YTO Express	March 2016	It was listed at a price of 17.5 billion yuan with with the backdoor of Dalian Dayang Trands Co., Ltd.
SF. Express	May 2016	It was listed at a price of 43.3 billion yuan with the backdoor of Dingtai
Yunda Express	July 2016	It was listed at a price of 18 billion yuan with the backdoor of Xinhai Shares
ZTO Express	October 2016	It was listed on the New York Stock Exchange and raised a total of $1.406 billion U.S. dollars
TTK Express	January 2017	Suning Commerce Group Co., Ltd acquired 100% shares of TTK Express for 4.25 billion yuan, which was not listed
Deppon Logistics	January 2018	It was listed on the Shanghai Stock Exchange

Source http://finance.eastmoney.com/news/1349,20151202571373622.html

3.2 Status Quo of International Express Delivery Giants

The express industries in developed countries, mainly the United States, Japan, and Germany, have experienced a long way of development and have possessed complete operation models. The industry was born in the United States, so what is happening in the U.S. is representative. Express delivery industry was born in Japan around the same time it was in China. But Japan is economically stronger, and the industry has developed more rapidly with a unique operation model. With the continuing improvement of China's comprehensive strength, many foreign-funded enterprises have been drawn to enter the market. They have set up joint ventures with domestic express delivery companies in China so as to achieve a localized strategic model.

3.2.1 United Parcel Service (UPS)

3.2.1.1 Development of UPS

United Parcel Service (UPS) was established in 1907 and headquartered in Atlanta, California, USA, which is the largest express delivery company in the world. UPS always takes the concept of building a comprehensive service company as its concept, actively participates in business development at home and abroad, and plays an indispensable role in business development. UPS has developed rapidly in air transportation. It took only 30 years to set up the UPS internal transportation fleet from the initial establishment to the expansion of large-scale services offered in various states in America.

UPS ranked the 138th among Fortune 500 enterprises in 2018. In 2018, UPS's operating revenue was $ 71,861 million, an increase of 9.1% over 2017. Figure 3.4 shows the rising operating revenue of UPS from 2009 to 2018, which grew slowly and incrementally.

The aviation distribution center is responsible for the transfer of aircraft and the sorting and processing of packages while the air navigation airport is responsible for the container transshipment and loading and unloading of parcels delivered by flight between each processing center and the airport within the radiation range. The air shipping center is responsible for the sorting of imported and exported parcels. The sorted packages would be transported to the airport by vehicles. The three kinds of centers form the UPS aviation network in the United States.

The key to UPS's success is its fleet and the use of IT in the transportation of goods. UPS has also developed DIAD that can offer package information of location to customers at any time. UPS provides customers with convenient services to keep track of the transportation details about packages through its official website.

In China, UPS mainly provides corresponding solutions according to the actual situation of its business partners. Initially, it entered China to develop its express delivery business through cooperation with Sinotrans Limited. With the growing

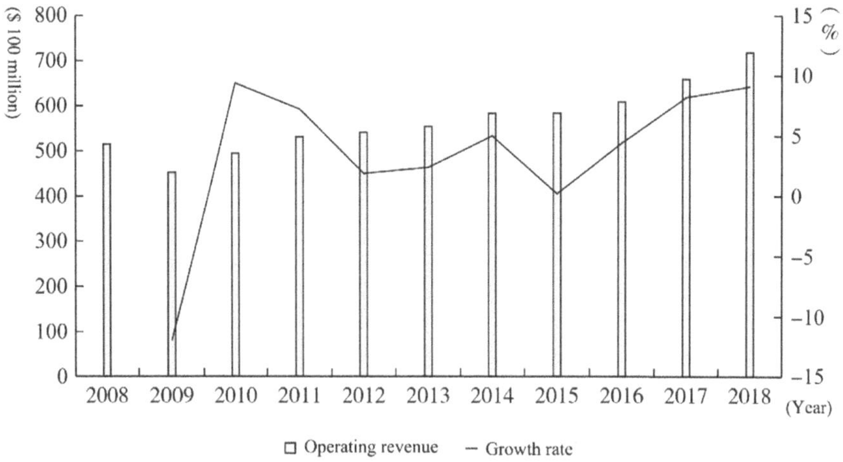

Fig. 3.4 The Operating revenue of UPS from 2008 to 2018. *Source* http://www.fortunechina.com/global1500/143/2018

demand of domestic and international express delivery from companies in China, It occurred to UPS to run domestic express delivery in China in 2005. Then it worked on the idea until it became true. At present, UPS sets up its network structure as a multi-hub-and-spoke network with three international air transshipment centers in Shanghai, Shenzhen, and Hong Kong respectively.

With the rapid development of the economy, the freight charges for different transportation approaches by UPS have also increased. According to official news, the UPS freight charges has increased by an average of 4.9% since December 24, 2017. UPS claims surcharges based on the weight of parcels. For example, if the parcel exceeds 70 pounds, it will be charged an additional fee of $19 in the U.S., and the surcharge for large packages delivered to U.S. residents from all over the world will be $90. This regulation has been in effect since July 8, 2018.

3.2.1.2 Development and Status Quo of UPS in China

(1) The Development of UPS in China

UPS entered China in 1988. In over 30 years, UPS currently holds business licenses in 33 cities in China and 252 business facilities It has established international transshipment centers in Shanghai, Shenzhen, and Hong Kong. The course of its development in China is shown in Table 3.3.

(2) Status Quo of UPS in China

China is the most important overseas market for UPS. There are 15 transshipment centers worldwide, three of which were established in China—Shanghai International

Table 3.3 The development of UPS in China

Time	Specific event
1988	UPS has began its express business in China through its first cooperation with a Chinese international trade transportation company
1996	UPS and Chinese international trade transportation company established a joint venture Sinotrans Beijing Airlines UPS International Express Co., Ltd. and established four offices in Beijing, Guangzhou, Shanghai and Shenzhen
2001	In January, UPS shipping freighter obtained the right of direct flight to China. In April of the year, UPS's first freighter flew directly to Beijing from the United States. The freight plane of UPS can fly directly to China, which marks direct express service between China and the United States started
January 2003	UPS began to cooperate with Yangtze River Express. Yangtze River Express provides flight inter-modal services, such as cargo transportation between the transshipment center established by UPS in Shanghai Pudong Airport and 4 major cities– Beijing, Xiamen, Guangzhou and Qingdao
December 2004	UPS Express has obtained the independent operation right of international express business in 23 regions of China through the acquisition of Sinotrans Beijing Airlines UPS International Express Co., Ltd., and its transportation services account for 80% of China's international trade
July 2005	UPS started to establish an international aviation hub in Shanghai while developing express services in China, and established direct flight transportation services from the United States to Guangzhou
2006	UPS officially launched the direct flight service from Shanghai to Europe and added three flights from Shanghai to the United States and one flight from Qingdao to Incheon, South Korea
2008	The international transshipment center established by UPS at Shanghai Pudong International Airport was put into use
May 2010	UPS's Shenzhen Air Transit Center at Shenzhen Bao'an Airport was officially operated
October 2012	UPS has launched a route from Zhengzhou to Anchorage International Airport in the United States. UPS developed express business makes Zhengzhou as a transshipment and distribution center for international and China's goods
August 2014	Since 2012, UPS has started express business within China. At that time, UPS has licenses in 33 cities of China
2015–2017	During the two upgrades of China–Europe railway service respectively, UPS used multimodal freight solutions to provide customers with FCL and LCL services, and established 6 stations in Zhengzhou, Chengdu, Changsha, Chongqing, Suzhou and Wuhan in China
May 2017	SF Express strategically cooperated with UPS to launch a new co-branded product, "SF-UPS Direct Shipping+", which firstly launched the service from China to the United States, and the transit time was stable within 7 working days
May 2018	SF Express and UPS jointly launched an upgraded version of the "Standard Express+" product named "SF-UPS Standard Express+". In Chinese mainland, SF Express provides a one-hour door-to-door pickup service. International transportation and delivery are arranged by UPSPackages can be shipped from China to Europe in as little as three days

Source https://www.ups.com/cn/zh/about.page

Air Transit Center, Shenzhen Asia Pacific Air Transit Center and Hong Kong Air Transit Center. UPS provides international and domestic express delivery business in China, and the former is the stress. About 200 flights per week provide transportation services for both domestic and international customers in China. UPS uses solid aviation power to link China closely with the global market.

UPS currently provides express services except for letters delivery business in 33 cities in China. It has deployed a large number of outlets in China and the service network covers more than 330 cities in China. It has also established a relatively complete service network and infrastructure. However, there are some disadvantages for UPS to run delivery business in China. First, network resources are insufficient for it, and there are limited transit centers, outlets, and vehicles. Secondly, compared with the low price of Chinese express delivery enterprises, the high price charged by UPS is disadvantageous and the publicity is not enough. What's worse, it is not a well-known brand for Chinese people. Therefore, UPS develops in a slow pace in China.

3.2.2 *FedEx*

3.2.2.1 Development of FedEx

Founded in 1937, FedEx is the fastest-growing company in the history of the U.S. express delivery industry. It is headquartered in Memphis. The idea of FedEx comes from the founder's vision of "overnight delivery" during his college life. FedEx aims at customer satisfaction. It promises to deliver in a specified time and give timely feedback on consumers' information, which is regarded as a guarantee for its development. It also proposes countermeasures for pricing based on whether the goods need urgent delivery or not. FedEx adopts the networked or closed-loop approaches to organize transportation for express delivery. The primary means of transportation is aircraft. The company itself possesses 677 self-owned airplanes and 44,000 professional trucks.

FedEx ranked the 155th among Fortune 500 enterprises in 2018, and the ranking rose up by 25 compared with that in 2017. The operating revenue reached 65,450 million US dollars in 2018, increased by 8.5% compared with that in 2017. The operating revenue (unit: million US dollars) of FedEx in the past 11 years is shown in Fig. 3.5. It can be seen from the figure that the operating revenue decreased from 2008 to 2010, but since 2011, it has been growing incrementally.

On the premise of shortening the customs clearance process of goods, FedEx developed an electronic system in 1993 that can inform the customs clearance procedures related to the package in advance before the delivery of packages meets the specified time. In 2010, FedEx developed a sensing device that can meet higher requirements of tracking the exact location of the goods anytime and anywhere and for some particular drugs which requires cryogenical storage for the effective medicinal use so as to consolidate the cooperation with pharmaceutical enterprises, FedEx

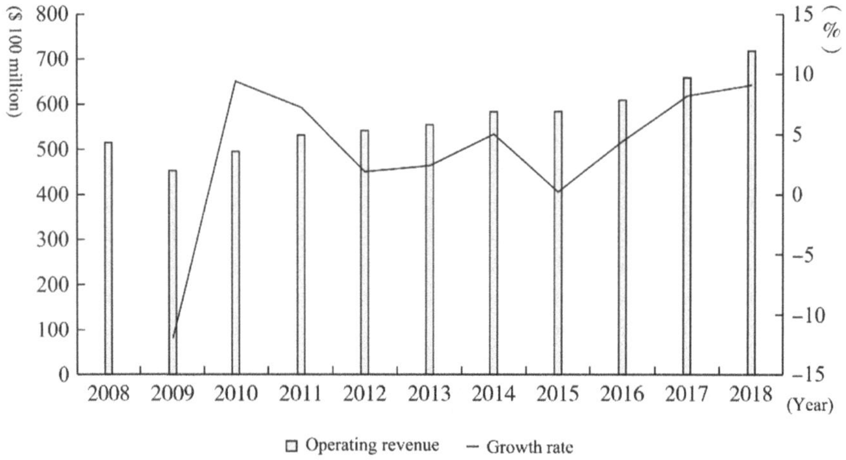

Fig. 3.5 The operating revenue of FedEx from 2008 to 2018. *Source* http://www.fortunechina.com/global500/200/2018

has launched an official download platform for clients, which provides services such as inquiring about cargo information and offering bills, and telephone services are also ready to bring convenience to clients. In 2016, FedEx acquired TNT Express, one of the world's largest express delivery companies providing road transportation and air freight services in Europe, the Middle East, Africa, Asia Pacific, and Americas.

FedEx believes that China's express delivery market has excellent prospects for its development through careful inspection. In 2007, it provided fast and convenient courier services for many cities and regions. Consumers in some areas could even get packages on the same day or the next day. In order to do better and faster in the future, FedEx has invested more funds in the planning and construction of its transmission network in China. A case in point is the further expansion of the infrastructure in its Shanghai Station. FedEx is premised on meeting the needs of Chinese consumers. In 2013, in order to meet the requirements of international cargo transportation, it completed the expansion of the port operation center in Beijing. According to the official news of FedEx, on January 1, 2018, it raised the delivery price of goods.

3.2.2.2 Development and Status Quo of FedEx in China

(1) The Development of FedEx in China

FedEx Express seized the opportunity of the rising of the express market in China and started its business here in 1984. Having developed for over 30 years, the express business of FedEx has covered 28 provinces, municipalities, and autonomous regions in China while its service network has laid out in more than 400 cities. FedEx has 78 branches in China, and the cities where these 78 branches are located all set up delivery points with more than 9000 employees. In China, FedEx has provides services in five airports of Beijing Capital, Shanghai Pudong, Shenzhen Baoan, Guangzhou Baiyun, Hangzhou Xiaoshan, and there are also large-scale port operation centers in Beijing, Shanghai, Shenzhen, and other cities. And its development history in China is shown in Table 3.4.

(2) Status Quo of FedEX in China

FedEx connects the service network in China with its global network. Currently, FedEx operates more than 220 flights per week in China, including over 160 weekly flights to and from the FedEx Asia Pacific Transit Center in Guangzhou. Enterprises in China has driven the booming development of express delivery enterprises around the country. FedEx's domestic express business in China is under pressure from Chinese express delivery enterprises. Express Mail Service represents the state-owned enterprises while the representatives of private enterprises are SF ZTO, YTO and Yunda Express. All of the above account for 90% of the express market in China and have extensive service networks covering across the country. Due to the high cost of the express delivery industry and the difficulty in networks development in China, obviously FedEx has a number of disadvantages in developing Chinese market. Firstly, the networks construction is sluggish, and there are less outlets. In addition, FedEx mainly operates international express delivery business in China, targeting high-end customers with strict requirements for timeliness and reliability. Therefore, it charges more. Compared with domestic express delivery enterprises, Price is not a competitive advantage for FedEx to run its business in China. Secondly, FedEx has not formulated a tailored operation mode and developing strategy regarding the local characteristics of China As a result, FedEx is relatively passive in its developing in China.

3.2.3 Deutsche Post (DHL)

DHL is a subsidiary of Deutsche Post DHL, a world-renowned postal and logistics group. The enterprise's name consists of the initials of the surnames of the three founders, Dalsey, Hillblom and Lynn. In 1969, they opened the first express delivery route from San Francisco to Honolulu. DHL is committed to establishing a new transport network that provides global door-to-door express services. In the mid and late 1970s, DHL continued to extend routes to South America, the Middle East and Africa (Lu, 2010).

DHL Air Freight express delivery company accommodates the acquired Airborne Express and Exel plc, forming one of the most convenient and widely distributed transport networks that can deliver goods to 220 countries and 120,000 regions around the world.

DHL started the shipping business in China on January 26, 2007, and has established the fastest and most comprehensive express service integrated network in

Table 3.4 The development of FedEx in China

Time	Specific Event
1984	FedEx cooperated with Chinese agents to carry out express business in China through commercial flights
1995	FedEx acquired Evergreen International Airlines, which has direct flights between China and the United States. In the same year, FedEx began to cooperate with China Sinotrans Co., Ltd. and established a joint venture to began its business in China
January 1996	The Civil Aviation Administration of China approved FedEx's application for air rights to China, and approved all-cargo air route between China and the United States
1997–1999	FedEx and Sinotrans Co., Ltd. terminated their cooperation. Two years later, they established a joint venture with Tianjin Datian W. Group. They used Datian's transportation network and infrastructure in China to expand express business within China
April 2006	FedEx acquired its express delivery network in China through the acquisition of Tianjin Datian W. Group, becoming the first foreign-funded express delivery enterprise in China, and began to operate express business within China
October 2008	FedEx is the first express delivery enterprise to launch a limited-time service in China, and has successfully established its strong operating network through the express services of "One-Day Delivery to Asia" and "One-Day Delivery to North America"
February 2009	The Asia–Pacific Transshipment Center established by FedEx at Guangzhou Baiyun International Airport has been put into use, which is FedEx's largest transshipment center at other country
February 2010	With the operation of FedEx's Boeing 777 cargo aircraft in Shanghai, China, it creates a point-to-point connection between Shanghai and Memphis, a supercenter of operations, thereby enhancing FedEx's international transportation capacity. In August and December of the same year, the expansion of the largest international express ground operation station in Shanghai and the operation of the second Boeing 777 freighter were completed in just 4 months
January 2011	FeEx has launched a new direct flight connecting India to the Asia–Pacific transshipment center in Guangzhou, China
June 2012	FedEx completed the expansive construction of the operating center at Shanghai Pudong International Airport
November 2012	FedEx put into operation of the fifth international express ground operating station in Nanshan District, Shenzhen
April 2013	FedEx has established its China highway transit centre in Wuhan, Wuhan beacme the national highway transit hub. It undertakes the cargo distribution and trans-shipment service of 8 main trunk highways and 16 round-trip from Wuhan to Xi'an, Zhengzhou, Changsha, Nanchang, Shanghai, Chongqing, Chengdu and Guangzhou. In July of the same year, FedEx completed the expansive construction of the operating center at Beijing Capital International Airport

(continued)

Table 3.4 (continued)

Time	Specific Event
January 2014	FedEx set up China Business Subregional Headquarters in Shanghai
April 2018	FedEx completed the connection business between FedEx Asia Pacific Transit Center in Guangzhou, China and FedEx Global Super Transit Center in Memphis, USA

Source http://www.fedex.com/zh-cn/about/history.html

China. According to the latest statistics, DHL operates in more than 220 countries and regions, employs above 285,000 people, builds over 450 warehouses and distribution stations, and transports at least 1.5 billion pieces of goods annually.

Among Fortune 500 enterprises in 2018, DHL ranked the 119th, stepping down 2 rankings compared with that in 2017. Its operating revenue reached 75,000.9 million euros in 2018, increasing by 7.2% compared with that in 2017. Due to the impact of the financial crisis, it can be seen from the figure in Fig. 3.6 that DHL suffered a huge loss from 2008 to 2009, and its business income decreased by 15.19%. And from 2009 to 2018, its operating revenue grew either steadily or decreased a little bit.

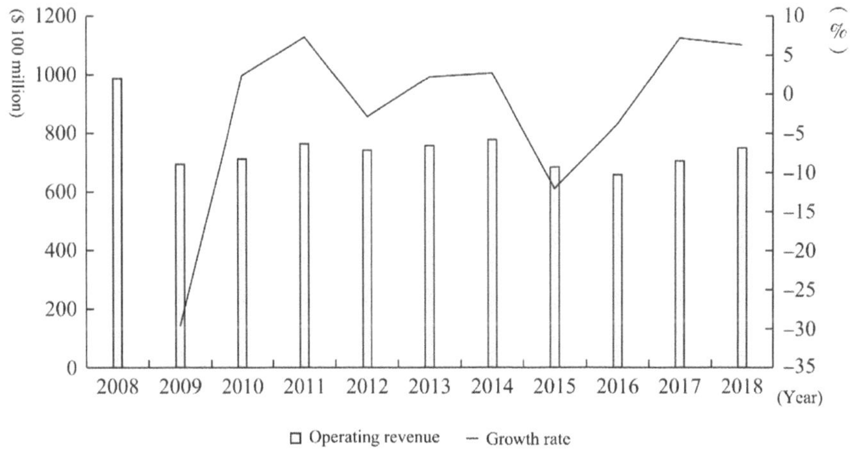

Fig. 3.6 The operating revenue of DHL from 2008 to 2018. *Source* http://www.fortunechina.com/global500/54/2018

3.2.4 *Japan Post*

Japan Post was founded in 1871, and the Japan Post Group was established on April 1, 2003, aiming at providing consumers with comprehensive, convenient and satisfactory freight and mail delivery services as well as banking services. Japan Post is a state-owned company that is financially independent. On November 4, 2015, Japan Post was listed on the stock market.

Japan Post faces brutal competition, especially with Yamato Transportation and Sagawa Express. According to statistics, in 2013, Yamato Transportation accounted for 36.3% of the market as the No.1 in the industry; Sagawa Express 33.9% as No. 2; and Japan Post 11.8% as No. 3. In 2014, the market share of Japan Post increased to 13% owing to its express delivery business.

In February 2015, a subsidiary of Japan Post Group acquired TOLL at A$ 6.5 billion. Japan Post has established more than 1200 business outlets by learning the advanced technologies and integrating the existing network resources obtained from the acquired company. In April of the same year, Japan Post formulated a medium-term business plan based on the group's status quo and long-term planning, setting precise goals for developing postal services.

Japan Post had launched a new service that allowed a courier to collect the recipient's response at the same time a mail is delivered, and deliver that to the sender of the original mail. The new service was available from July 1, 2015 until the end of June 2016, and its target clients were those who needed to send out letters in bulk.

On October 2, 2015, STO Express Japan Branch "Japan STO" and Japan Post Co., Ltd. signed a comprehensive strategic cooperation agreement. The two sides work together to provide international logistics services such as cross-border E-commerce customs clearance and "direct delivery of tracking number" for Chinese cross-border e-commerce customers in Japan. It only takes 5–7 days to complete the whole process.

Among Fortune 500 enterprises in 2018, Japan Post ranked 45th, down by 12 compared with 2017, and its operating revenue in 2018 fell again to 115,220.5 million US dollars, decreasing by 1.2% from 2017. The operating revenue (unit: million US dollars) of Japan Post in the past 11 years is shown in Fig. 3.7. And in 2010, the revenue of Japan Post reached 203,958.1 million US dollars. Since 2011, it has shown a falling trend. Until 2014, it began to increase slowly, but was still lower than that in the previous peak years.

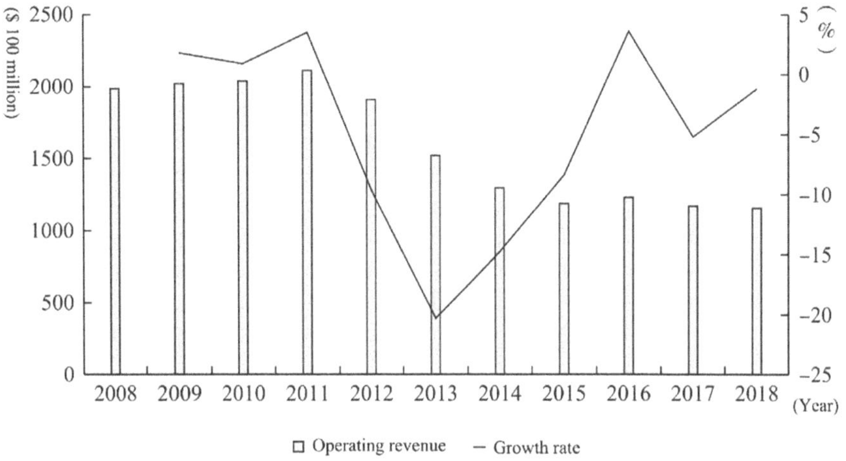

Fig. 3.7 The operating revenue of Japan Post from 2008 to 2018. *Source* http://www.fortunechina.com/global500/11/2018

3.3 Status Quo of the State-Owned Express Delivery Companies in China

Since the 1980s, China's economy has developed rapidly. The express delivery industry emerged with the continuous expansion of international communication and trade. State-owned express delivery enterprises such as China Post, China Railway Express and Civil Aviation Express have developed rapidly owing to their monopoly resources.

3.3.1 China Postal Express (EMS)

3.3.1.1 Development of EMS

China Post is a wholly state-owned company managed by central government, the full name of which is China Post Group Co., which is mainly responsible for postal services in the Chinese mainland. On December 3, 1985, China Post established its subsidiary China Express Service company, and it is the first express delivery enterprise in the country. On August 1, 2001, the domestic parcel delivery business was launched, providing land express delivery services. As an enterprise with the largest express delivery network resources, China Post assumes the function of providing universal services.

China Post's express delivery business operates in more than 200 countries and regions worldwide, including 31 provinces, municipalities, and autonomous regions, and also Hong Kong S.A.R, Macao S.A.R, and Taiwan, with more than 45,000 service outlets. China Post is striving to create a world-class postal land transport network, reform the express parcel business, adjust the layout of the land transport network, accelerate the transformation process of the land transport network, and reconfigure the construction and functions of outlets. In 2015, China Post formulated the operating standards for package delivery and standardized regulations on collecting, sorting, transporting and delivering parcels, which greatly improved the overall operational efficiency of express delivery. To fully implement unified operation, the branches of China Post in each province completed the optimization of the provincial network.

China Post's huge transportation network includes 668 first-level arterial roads and 2916 second-level arterial roads. China Post has formulated mechanisms and methods of operational management in the land transportation networks to improve its quality and ability.

There are 151 sets of sorting equipment in the network, including 70 sets of parcel sorters, 71 sets of letter sorters, and 10 sets of flat sorters. In 2015, China Post invested 920 million yuan to completely constructed the streamlined process of mail processing centers and reform its operational method, and update the parcel sorters of 19 mail processing centers. After streamlined transformation and the application of a new type of parcel sorter, the whole network implemented a new production process of bulk and pipeline network transportation, canceled the packaging method of pouches, and implemented operations such as the truck being able to go at any time and no backlog of packages. Its handle record of packages reached 4.01 million in one day, and the double-deck sorter handles more than 470,000 pieces daily. In addition, strengthened processing capacity of 31 secondary central bureaus and the technological transformation of processing sites in prefecture-level cities were launched, which effectively strengthened the processing capacity in prefecture-level cities.

The informatization level has been significantly improved. A total of 135 service devices were newly added. And the efficiency of the main system was significantly improved for customers feedback and production inquiry, which greatly ensure the smooth operation of key projects such as Jisuxian, large packages, and government courier services. They also helped clients meet the demands of business peak in "double 11" efficiently and smoothly. There have been an improved security management system and a national central quarantine zone for internet security, leading to the integrity rate of system operation reached 99.7% throughout the year.

3.3.1.2 Network Structure

China Postal Express and Logistics adopts a comprehensive radiation-type network structure, presented in the form of a top-to-bottom network structure. It is also called Centralized Radial network Structure which combines several basic models. China

Postal Express and Logistics has a strong express service network through the outlets established in the early stage around the nation. According to the spatial distribution, the postal express service network is composed of seven regional sections: Southwestern China Post, Northern China Post, Southeastern China Post in Coast Areas, Northeastern China Post, Eastern China Post, Central China Post, and Northwestern China Post. On this basis, node cities with hub functions are determined in the seven sections, which are the key nodes of the national postal network. These cities have the characteristics of comprehensive transportation hubs and are mainly responsible for cargo distribution and transshipment, followed by provincial-level cities or some cities on the arterial roads as sub-level hub cities with a medium-sized postal processing center in the network. Finally, nodes in cities or smaller administrative units are where fulfillment centers are located.

Although China Postal Express and Logistics has strong network resources across the country, its businesses were not balanced. Express outlets are dense in the eastern regions and sparse in the western areas. The overall trend is decreasing from the east to the west. The quality of delivery services also needs to improve from coastal areas to inner land. Although the distribution network in Northern China is relatively complete, the overall network layout is unbalanced. It is suggested that the company strengthen the construction of network nodes, optimize the existing express delivery network, and balance the general layout of service network.

3.3.2 China Railway Express (CRE)

3.3.2.1 Development of CRE

In 1993, China Railway established the express department only for the transportation of small goods. In April 1997, China Railway Express Co., Ltd. was established in Beijing, the capital of China. In 2004, the company changed into a Limited Liability company. It is a modern integrated logistics group covering e-commerce services, warehousing, truck-load transportation, finance, cold chain, customs declaration and inspection agency, trade, international express and other businesses. At present, it is an exclusive modern integrated 5A logistics enterprise with whole industry chain in China.

With the goal of making the company bigger and stronger, China Railway Express has formed a transportation network of 1000 luggage trucks in major large and medium-sized cities across the country and set up 2030 business offices in more than 670 cities so as to provide services with security and punctuality (Wang, 2015). It is the most extensive professional express delivery network in China. Thanks to its advantageous network and the characteristics of safe and fast railway transportation, the company possesses railway luggage transportation resources across the country. China Railway Express also provides integrated supply chain services for 5100 Tibet Spring. It has created many jobs for people live in Xizang and eased unemployment. In order to ensure the timely delivery of parcels to customers, China

Railway Express operates 42 express trains every day and strictly follows the regular transportation schedule every day. The annual throughput exceeds 13 million tons, forming a complete domestic railway express transportation network.

In 2014, CRE launched the high-speed rail express service for the transportation of goods among economically developed cities, such as Beijing and Shanghai. In June 2017, during the "Belt and Road" International Cooperation Summit Forum, China Railway Express cooperated with Russian Railway and its logistics company, establishing a Sino-Russian cross-border e-commerce logistics platform.

3.3.2.2 Network Structure

The transportation channel of China Railway Express is composed of railway transportation lines with trains, road transportation arterial roads and air transportation lines. And the main transportation means are through railway, highway and airlines. Road transportation can deliver goods in a door-to-door way so as to compensate for what railway cannot do. The services of China Railway Express is mainly divided into high-speed rail express and ordinary express. The express delivery network of high-speed rail express mainly relies on high-speed rail transportation, while the express delivery network of ordinary express mainly adopts ordinary railway, road and air transportation.

The express delivery network of China Railway Express falls into two tiers of nodes, and it is divided into seven blocks: Northwest China, Southwest China, Central China, South China, Northeast China, North China, and East China. The first tier is regional transit centers which are the highest-level node in the network and the largest distribution hub centers of China Railway Express. Each of them is responsible for transshipment in the areas close to it. Regional transit centers are directly interconnected through "point-to-point" methods, forming the backbone network of China Railway Express. The second level is urban transit centers which undertake transshipment between regional transit centers and service outlets.

The express delivery network structure of China Railway Express is a hub-and-spoke one with multiple hubs, as shown in Fig. 3.8 (Fan, 2007). The main frame is composed of regional transit centers across the country. The distribution network is composed of city transit centers and service outlets. Terminal pickup and delivery network consists of customers and service outlets. The hub-and-spoke express delivery network with a multiple-hub structure can quickly collect express packages from different places. Packages are collected from service outlets and passed to city transit centers, and then quickly transported to regional transit centers. This aggregation method can greatly reduce the operation time.

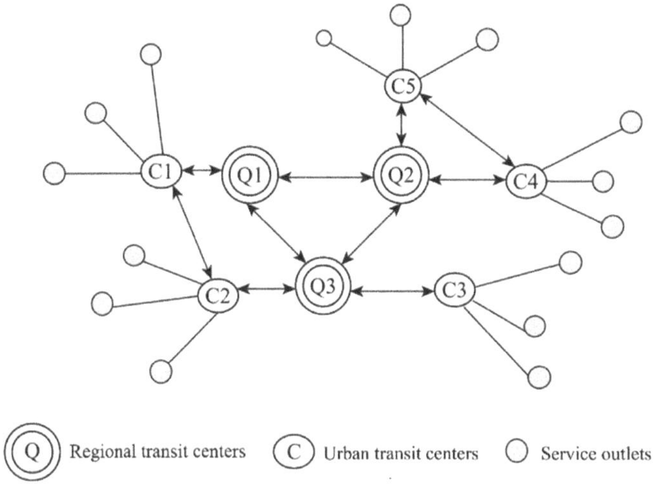

Fig. 3.8 The "hub-and-spoke one with multiple hubs" mode in the express delivery network structure of China Railway Express. *Source* The Author

3.3.3 *Civil Aviation Express (CAE)*

In 1996, Civil Aviation Express Co., Ltd. affiliated with China National Aviation Corporation was established. For 13 consecutive years CAE has been included in the list of "China's 500 Most Valuable Brands". In 2017, the brand value reached 12.521 billion yuan. Civil Aviation Express takes trustworthiness and customer first as its business philosophy, strengthening brand, talents and transportation network construction.

Civil Aviation Express has opened international direct freight from Guangzhou and Shenzhen to Hong Kong and land transportation network centers in the Pearl River Delta, Yangtze River Delta, and Northern China. Owing to the unique advantages of possessing 1578 domestic routes and 302 international routes, it has established 363 outlets in large and medium-sized cities in China and gradually formed four production and operation distribution centers in Beijing, Guangzhou, Shanghai and Chengdu to provide safe, convenient, punctual, trustworthy and satisfactory services for customers (Li, 2015).

Civil Aviation Express promotes comprehensive logistics solutions by means of advanced management information, with a combination of information technology. It keeps striving to consistently provide high-quality logistics services for customers and enhancing economic benefits.

In 2017, Civil Aviation Express was rated as the KPI bench-marking enterprise in the automotive logistics industry and the best service enterprise in pharmaceutical cold chain logistics.

3.4 Status Quo of the Private Express Delivery Enterprises in China

In the early 1990s, the private express delivery enterprises in China began to develop. In order to seize market share, many of them adopted low-price strategies and set up a lot of outlets to enter the express delivery market and develop rapidly. Currently, private express delivery enterprises have a market share of close to 70% in the Chinese mainland. This chapter categorized private express delivery enterprises into three types: the first is direct operating mode adopted by SF Express. Services are provided by the directly-operated outlets. Prices are high while delivery is quick. Companies possess advanced backstage technology; the second is the self-operated logistics mode like JD Logistics. Its distribution efficiency is high and services are of good quality; the third is the franchise mode adopted by STO, YTO, ZTO, Yunda and other express delivery enterprises. In this mode, the lowest costs make it possible to offer the lowest price. Services are acceptable in terms of speed and reliability, and enterprises with this mode compete fiercely with one another.

3.4.1 SF Express

3.4.1.1 Development of SF Express

SF Express was established on March 26, 1993. SF Express has now offered same-day arrival air express service in some cities like Beijing, Shanghai, Guangzhou, and Hangzhou. And SF Express provides night pick-up services for consumers in Beijing, Tianjin, and Shandong Province.

SF Express adopts a top-down and partitioned organizational management mode (Guo, 2013). Its organizational structure is composed of headquarters, regions, divisions, and outlets. SF Express is headquartered in Shenzhen and has five market regions—South China, East China, North China, Central China and Southeast China. With overseas market, there are six operating centers.

In terms of information technology application, SF Express has cooperated with internationally renowned enterprises such as ORACLE to improve the automation level of equipments. It has established an information system with functions such as express delivery transfer, information monitoring and express query. Since its establishment, SF Express has focused on service quality, strengthened infrastructure construction, and established large express service organizations and service networks for information gathering, market development, logistics, and distribution at home and abroad.

Consistently strengthening express services, SF Express adheres to take customers needs as the core of its business, actively expands diversified services, such as launching integrated supply chain solutions for customers with different needs like food and medicine, and provides comprehensive financial services of payment, financing, wealth management and value insurance. At the same time, relying on its strong logistics advantages, SF launched SF Optimization. It provides customers with high-quality life services and creates a high-quality life experience. All along, SF Express has been constantly innovating and working hard to meet the needs of different customers.

In early 2015, SF Overseas Shopping made its official debut. SF Express has successively launched "Overseas Shopping SF Express" and "Global SF Express" to open the European market on a full scale. SF Express now has two major shopping platforms, "SF Optimization" and "SF Overseas Shopping". What's more, SF Express can realize both express delivery and online shopping on its APP.

SF Express officially changed its name to SF Holding on February 24, 2016. There are 6 distribution centers (Beijing, Shanghai, Hangzhou, Guangzhou, Shenzhen, and Xiamen) across the country, of which Shenzhen center is the largest. Goods are transported between the six distribution centers by air. On December 12 of the same year, SF Express obtained a permit from China Securities Regulatory Commission (CSRC) and was allowed to be listed on the A-share market.

In the early morning of June 1, 2017, SF Express announced that it would close the data interface to Cainiao network Technology Co. From noon on June 3, 2017, SF Express resumed full business cooperation and data transmission with Cainiao. On November 20, 2017, the auction of three Boeing 747 airplanes was officially launched on Alibaba's auction platform, with SF Airlines Co. Ltd. auctioning off two of them for 160.7808 million yuan and 160.386 million yuan respectively, at a total cost of over 320 million yuan.

In March 2018, CAAC East China Regional Administration issued the first domestic UAV (Unmanned Aerial Vehicle) aviation operation (pilot) license to Jiangxi Fengyu Shuntu Technology Co. Ltd., a subsidiary of SF Express. On March 26, 2018, the BrandZ™ 2018 Top 100 Most Valuable Chinese Brands list was released, with SF Express ranked 11th.

3.4.1.2 Network Structure

The transport network of SF Express started to develop from Guangdong, and gradually expanded to the Pearl River Delta, East China, Central China, and North China. Its express outlets have extended to major cities across the country. In 2009, SF Airlines Co. Ltd. started its business and became the first private enterprise in China with its all-cargo aircraft. And Hangzhou is the air transport hub centre of SF Express in China. Its distribution capacity enhances the delivery and air transport speed of express westward from East China, and drives the flow of express from North and South China to the Southwest region.

(1) The Spatial Structure of Aviation network

At present, 52 freighters of SF Express serve for 36 cities across the country. Considering market demand and factors of urban aviation service facilities, cities enjoying the services are mainly in the southeast coastal economic belt and central and western provincial capitals. SF Express has created a "dual hub and spoke" air transport network mode, which takes Shenzhen and Hangzhou as shipping hubs to radiate to major cities across the country to form a domestic airline network. The non-stop transport mode is adopted between Shenzhen and Hangzhou. The transport mode primarily based on night flights can effectively avoid the peak passengers flow to achieve the punctuality of "overnight arrival." The two major air transport hubs in Shenzhen and Hangzhou in the transport network have 5 fixed air flights per day. The transport frequency between the two hub centers and their radiation areas is flexibly arranged according to the actual traffic volume.

As depicted in Fig. 3.9, SF Express adopts a "dual hub and spoke" mode that combines the "point-to-point" and "hub-and-spoke" air transport network modes. The advantage of "point-to-point" travel is that there is no need to go through a hub to get a direct express connection between two places, which saves a lot of time. The disadvantage is that the frequency of aircraft needs to be increased to meet the demand and the cost is higher. The advantage of using the "hub-and-spoke" model is that it provides economies of scale; however, the downside is that it requires passing through aviation hubs and is not very time-efficient. SF Express has adopted a "dual-hub-and-spoke" air transport network mode, combining both methods to maximize time efficiency and achieve economic scale effects. Although this mode of operation has many advantages, further research is needed to determine the optimal frequency of flights for maximum economic efficiency.

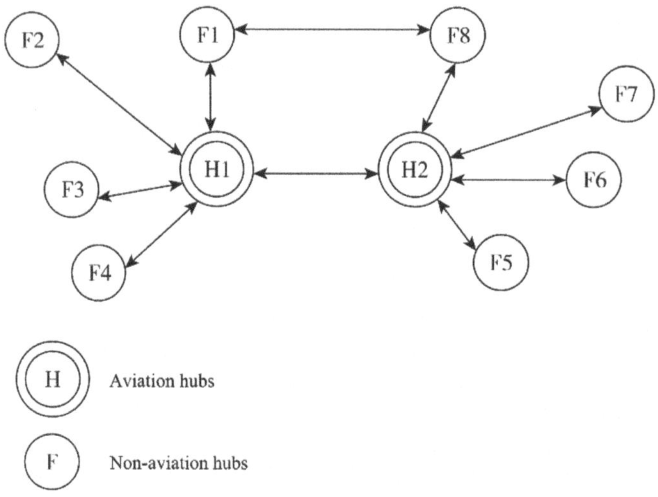

Fig. 3.9 The "dual hub-and-spoke" air transport mode of SF Express. *Source* The Author

(2) The Ground Network of SF Express

SF Express has created a "multi-hub hub-and-spoke" mode for its ground transport network, which links hub-and-distribution center cities. The hub and city transfer station are connected using a hub-and-spoke form that utilizes trunk corridors. Currently, SF Express has seven regional distribution networks in China, with Shenyang, Beijing, Shanghai, Guangzhou, Chengdu, Wuhan, and Xi'an as the hubs for Northeast China, North China, East China, South China, Southwest China, Central China, and Northwest China, respectively. These hub cities collect and distribute goods as the core of the regional express, and the goods are sorted and distributed to the city transit and stores. As is shown in Fig. 3.10.

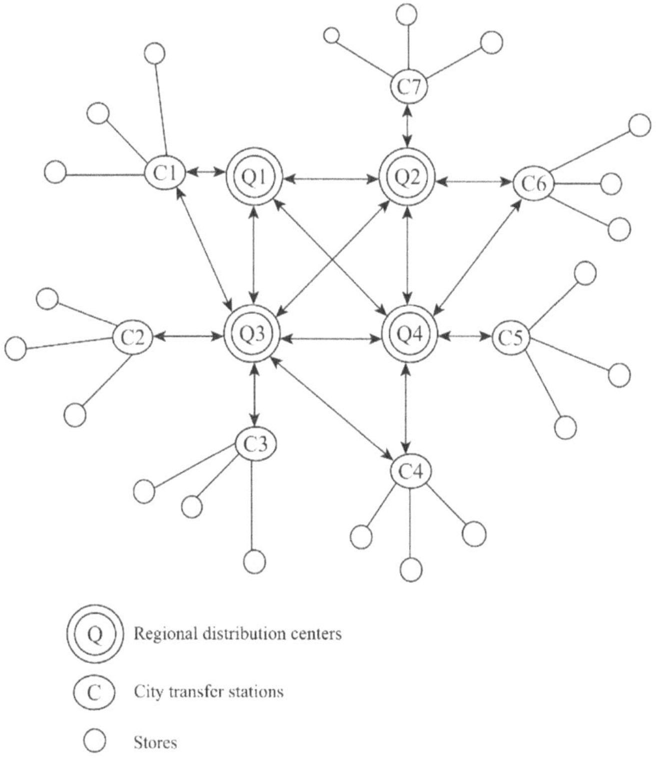

Fig. 3.10 The "Multi-hub hub-and-spoke" ground transport mode of SF Express. *Source* The Author

3.4.2 JD Logistics

3.4.2.1 Development of JD Logistic

Founded in Beijing on June 18th, 1998, JD Logistics offers a wide service area with 263 large warehouses, 6780 distribution stations and self-pickup sites, and 9 "Asia No. 1" large-scale intelligent logistics centers across the country.

In 2007, JD Logistics began constructing its own logistics system, which could arrange deliveries according to the customer's desired time, particularly for self-operated products, which were fast. For instance, if an order was placed before 11 pm, it would be delivered to the customer the following morning, or the customer would get it the same afternoon if the order was made before 11 am.

JD Logistics has implemented a self-crafted logistics strategy which has drastically improved the rate of transfer between warehouses, resulting in shortened delivery times to customers. Simultaneously, JD Logistics is excelling in its development while providing its partners with comprehensive product services, such as integrated supply chain solutions, merchant data services, and cross-border logistics services.

JD develops environmentally friendly materials through the use of technological innovation in order to create green logistics. It has the world's only six major logistics networks, and they use big data, artificial intelligence, and intelligent supply chain to accurately analyze product sales and stay up to date with all aspects of goods distribution in real time.

As JD's development has been successful, it has comprehensively promoted e-commerce poverty alleviation through the sale of local specialties. In 2016, the sales of poverty alleviation agricultural products reached nearly 10 billion yuan. With an efficient transportation network and perfect logistics infrastructure, JD has been able to reduce delivery time, cut costs, and increase operational efficiency due to its network, supply chain, and implementation capabilities.

3.4.2.2 Network Distribution

JD is a highly-acclaimed self-operated e-commerce company renowned for its fast delivery and excellent service. In 2007, it established a self-operated logistics system, followed by Shanghai's Haiyuanmai Logistics two years later. This led to the creation of a nationwide logistics and distribution system. In 2010, the "East China Logistics Warehousing Center" invested in and constructed by JD in Shanghai became the largest logistics warehousing center. JD Mall has set up six logistics centers in Beijing, Shanghai, Guangzhou, Chengdu, Shenyang and Xi'an, and has established urban distribution stations in more than 300 key cities around the six major logistics centers, forming a nationwide self-operated logistics distribution system. In addition, JD's logistics distribution system and facilities are constantly being improved

Table 3.5 The distribution of JD Logistics outlets

Province	Quantity	Province	Quantity	Province	Quantity
Beijing	403	Shandong	95	Guizhou	27
Jiangsu	226	Henan	87	Gansu	26
Guangdong	222	Hunan	85	Jilin	19
Hebei	142	Liaoning	77	Hainan	18
Zhejiang	130	Anhui	77	Heilongjiang	17
Shaanxi	124	Neimenggu	48	Ningxia	11
Shanghai	121	Fujian	46	Yunnan	9
Shanxi	115	Chongqing	44	Qinghai	9
Hubei	110	Tianjin	43	Xinjiang	3
Sichuan	100	Guangxi	35	Xizang	1

Source https://www.jdl.cn/unroute/map

to enhance the speed and quality of delivery, which now includes self-pickup cabinets, drones and speed delivery. Recently, JD's drones have been put into operation in Xi'an, Beijing and other places, while self-pickup cabinets are also widely distributed in various outlets. Furthermore, their next-day delivery service covers 248 cities across the country. According to the data provided by JD's official website, the number of distribution of JD outlets up to 2017 is shown in Table 3.5.

Figure 3.11 illustrates the division of JD across various provinces and regions in the country, with the corresponding number of outlets for each region. These include 751 outlets in North China, 695 outlets in East China, 113 outlets in Northeast China, 275 outlets in South China, 282 outlets in Central China, 181 outlets in Southwest China, and 173 outlets in Northwest China.

JD Express outlets cover all cities above the prefecture level in the country, particularly in the economically developed and flat areas of Beijing-Tianjin-Hebei, Jiangsu, Zhejiang and Shanghai, as well as the Yangtze and Pearl River Deltas; whereas the Yunnan-Guizhou Plateau, the northwest region, and the three eastern provinces are more sparsely-covered.

3.4.3 STO Express

3.4.3.1 Development of STO Express

In 1993, STO Express, one of the earliest domestic express delivery brands, was established. After more than ten years of development, it has formed a nationwide self-operated express delivery network, becoming the largest and most comprehensive private express delivery network in the domestic market. The company focuses on the development of national brands, using information technology as the basis

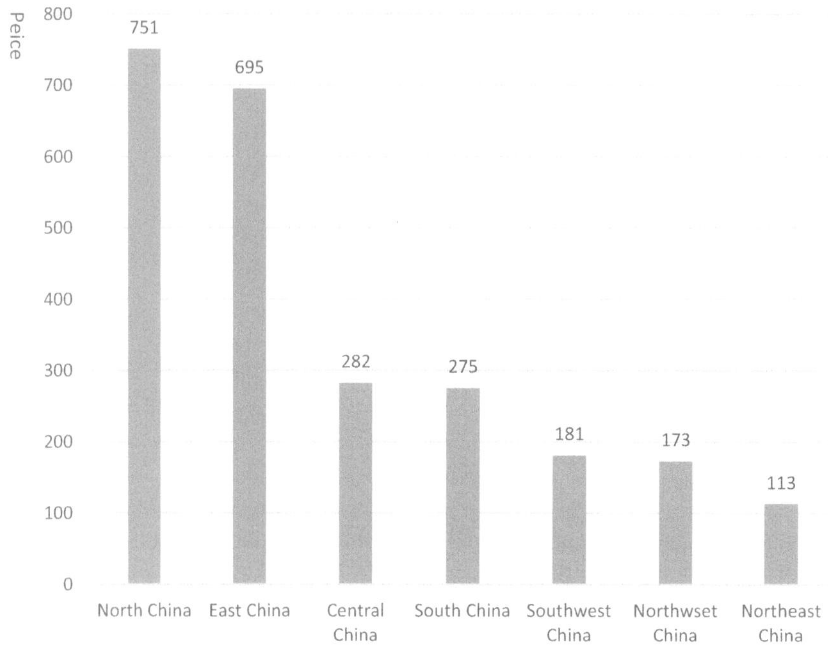

Fig. 3.11 The regional division of JD Express. *Source* https://www.jdl.cn/unroute/map

for innovation and growth, continuously improving its network operating system, and fully entering the field of e-commerce logistics due to the rapid development of the economy. STO directly manages the dialing and distribution centers and has its unique management method. It promotes the development of China's express delivery industry by providing professional services and strictly controlling quality.

STO Express has invested heavily in the research and development of the "STO E3 Express Software System Platform", which boasts powerful features such as order data collection, weighing, billing, and a customer complaint system. In the face of fierce competition, China's express delivery market is rapidly developing. Realizing that its existing express delivery services would gradually become obsolete in this competition, STO Express began to expand into new business areas such as third-party logistics, warehousing, payment collection, and international parcels to consolidate its foothold in the courier industry.

In 2017, STO Express invested heavily in automated sorting, scanning, weighing, and bubble counting equipment in order to bring smart logistics to fruition, particularly the application of the "Little Yellow People" sorting system in Yiwu, Zhengzhou, Tianjin, and Linyi, which made it an internet sensation for a while. This has added a remarkable invention to the technological innovations of China's express delivery industry. In terms of environmental protection, STO also launched environmentally friendly chip bags, which have several advantages over disposable woven bags, including being waterproof and wear-resistant, reducing damage, being cost-effective, being made of environment-friendly materials, and having unified specifications.

As of August 2017, STO Express has successfully implemented the use of environmentally-friendly bags in its network of transshipment centers. Every day, 180,000 of these bags are recycled, resulting in the conservation of 1.971 billion disposable woven bags each year. STO Express is putting its commitment to low-carbon environmental protection and green development into practice.

According to the latest official news, the company had 1899 independent outlets and branches, more than 15,000 township outlets, and 300,000 employees by the end of January 2018. This has resulted in a comprehensive and efficient self-operated express delivery network that spans the entire nation.

3.4.3.2 Network Distribution

STO Express outlets cover cities above prefecture-level across the nation, especially in Jiangsu, Zhejiang, and Shanghai, ensuring that no area is left without delivery service. All the outlets listed on their official website are regional company headquarters. According to the data provided by STO's official website in 2017, the distribution of STO Express outlets is shown in Table 3.6.

Figure 3.12 illustrates the division of STO in various provinces and regions across the country, as well as the number of outlets in each region. East China has 581, Central China has 379, Southwest China has 376, North China has 326, Northwest China has 314, South China has 190, Northeast China has 181, and Hong Kong and Macao have 2.

Due to the struggling economy and low population density in the remote western regions, the outlets have high fees, lengthy delivery times. Thus, the distribution

Table 3.6 Distribution of the number of STO Express outlets

Province	Quantity	Province	Quantity	Province	Quantity
Hubei	149	Guangxi	93	Chongqing	57
Hebei	143	Xinjiang	91	Shanghai	55
Henan	127	Jiangxi	90	Jilin	37
Yunnan	124	Heilongjiang	85	Qinghai	28
Sichuan	110	Guizhou	81	Ningxia	20
Shandong	108	Gansu	81	Tianjin	4
Shanxi	108	Jiangsu	79	Hainan	4
Hunan	103	Anhui	78	Xizang	4
Zhejiang	101	Fujian	70	Beijing	3
Shaanxi	94	Neimenggu	68	Hong Kong	1
Guangdong	93	Liaoning	59	Macao	1

Source http://www.sto.cn/Service/CustomerService?active_li=2&active_span=23

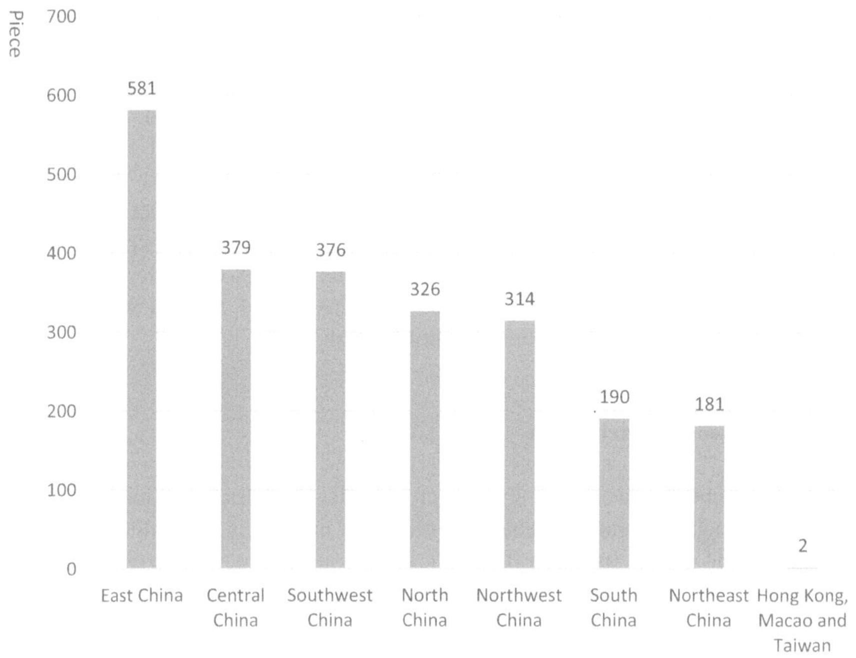

Fig. 3.12 The regional division of STO Express. *Source* http://www.sto.cn/Service/CustomerService?active_li=2&active_span=232&active_span=23

of outlets in remote western areas should be more extensive. In other words, the distribution of express outlets is also strongly linked to the transportation network. As economic and transportation hubs, coastal cities are densely populated and, consequently, the distribution of outlets is the densest in these areas.

3.4.4 YTO Express

3.4.4.1 Development of YTO Express

YTO Express was established on May 28th, 2000 and is an express service company that specializes in express, aviation, e-commerce, and other related businesses in China. After 20 years of growth and development, it has become a sizeable enterprise group that focuses on collectivization, branding, and networking. Its goal is "YTO Express - express delivery for Chinese" and its service tenet is to satisfy customer needs.

YTO was established in Shanghai, and as the express delivery industry rapidly developed in the domestic market, their business volume increased and their network grew, eventually expanding to the entire Yangtze River Delta region (including

Jiangsu Province, Zhejiang Province, etc.) before entering East and Central China, with their network distribution also expanding.

From a distinct standpoint, YTO Express has launched domestic services such as 12-h next morning delivery, 24-h next day delivery, 36-h next day morning delivery, signed return, payment collection, and other express services to meet customer needs and extend YTO's service network to a global level. Presently, YTO's business scope encompasses all townships in 31 provinces, municipalities, and autonomous regions across the country and more than 200 countries and regions around the world, with more than 45,000 business outlets.

At the end of 2010, Shanghai Yuantong Jiaolong Investment and Development (Group) Co., Ltd. was established. YTO was determined to expand sales, strengthen cooperation with others, develop e-commerce, and enhance service quality as it developed. This marked a firm step for YTO to look for opportunities in the competition and seek development.

In 2018, YTO began its B network business in nine second- and third-tier cities (Taizhou, Yangzhou, Jinhua, Wenzhou, Nanchang, Changsha, Guiyang, Dalian, and Changchun). By the end of the year, the business had expanded to more than 1000 cities nationwide. On March 26th, 2018, YTO Express ranked 36th on BrandZ™ 2018 Top 100 Most Valuable Chinese Brands list.

3.4.4.2 Network Distribution

As of December 31, 2017, YTO Express had 60 self-operated hub transshipment centers and 24,000+ terminal outlets nationwide, covering 31 provinces, autonomous regions, and municipalities; cities above the prefecture level had full coverage, and cities above the county level had a coverage rate of 93.9%. Table 3.7, provided by YTO's official website, displays the distribution of YTO Express outlets.

In Fig. 3.13, there are 387 outlets in South China, 923 in East China, 345 in Northeast China, 544 in Southwest China, and 54 in Northwest China, totaling to 349 outlets. There is only one outlet in Hong Kong, Macao, and Taiwan. The total number of outlets in the western region such as Xinjiang, Qinghai, Neimenggu, etc. is fewer than that of the other regions.

YTO's network coverage rate decreases from Central to East and West parts of the country; generally, the more developed the economy is, the higher the population density and more concentrated the outlet distribution. Economic level and population density are two important factors that affect the growth of express outlets.

Table 3.7 Distribution of the number of YTO Express outlets

Province	Quantity	Province	Quantity	Province	Quantity
Guangdong	237	Shaanxi	128	Xinjiang	94
Hebei	199	Hubei	118	Fujian	89
Sichuan	185	Liaoning	117	Chongqing	83
Zhejiang	182	Guizhou	116	Gansu	80
Shanxi	177	Heilongjiang	116	Tianjin	64
Shandong	172	Guangxi	114	Hainan	36
Henan	167	Jiangxi	114	Qinghai	26
Yunnan	153	Jilin	112	Ningxia	21
Hunan	135	Beijing	104	Xizang	7
Shanghai	134	Neimenggu	103	Hong Kong	1
Jiangsu	130	Anhui	102		

Source https://www.yto.net.cn/express/service/querysupport/branchsearch.html/

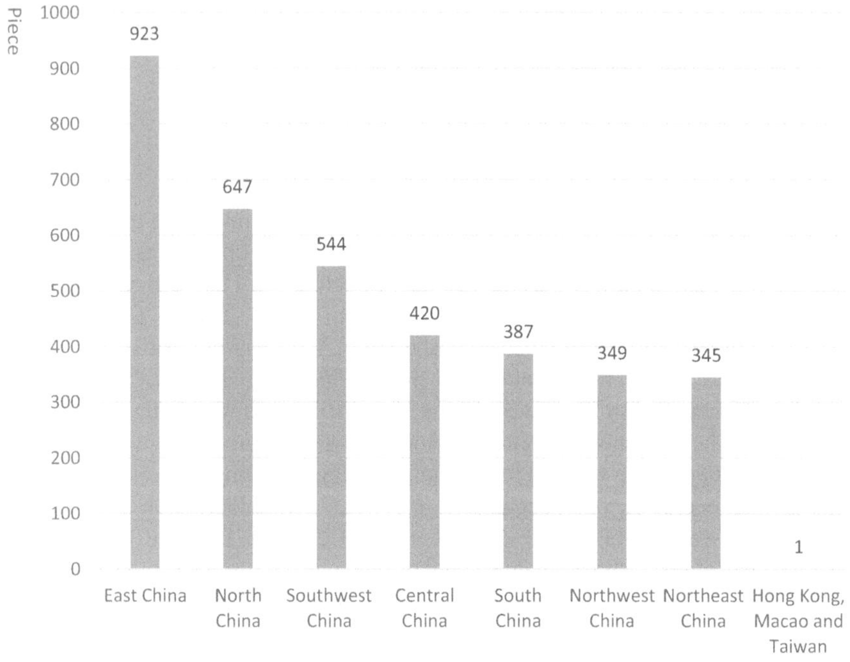

Fig. 3.13 The regional division of YTO Express. *Source* https://www.yto.net.cn/express/service/querysupport/branchsearch.html

3.4.5 *ZTO Express*

3.4.5.1 Development of ZTO Express

Established in May 8, 2002, ZTO Express is a large corporation that combines express delivery, logistics, e-commerce and printing businesses. The company has focused on building up its technical infrastructure, and in 2010 put in effect the "base of headquarters" management strategy. This strategy is based on the concept of "people-oriented, management with integrity", and has helped ZTO Express to grow and evolve consistently. Government policy has intensified the competition in the express delivery industry, which has in turn enabled ZTO Express to optimize its operations, innovate to expand its market and offer its customers safe, fast and high quality services.

In recent years, ZTO Express has been increasing its investment, resulting in a considerable amount of jobs and tax revenue, which has led to the sustainable and healthy growth of China's economy. It has invested 10 billion yuan in building large-scale distribution centers, e-commerce centers, etc., to enhance the transfer of goods. With the continuous opening of the Chinese market, international express delivery corporations have ventured into the express delivery market, causing profits to decline continually. Private companies, however, have been provided with opportunities due to the support of national policies and the board market. Amidst these opportunities and challenges, ZTO Express has strictly implemented time-limited delivery and standardized its operational procedures.

The express information system independently developed by ZTO has been successfully implemented. Furthermore, mechanized operations are employed for sorting and assigning tasks to improve distribution efficiency. ZTO Express is also dedicated to optimizing delivery routes and enhancing service quality through regular surveys and updates. Driven by substantial investments and tremendous efforts, ZTO Express has made tremendous progress. In 2014, the total volume of ZTO's business exceeded 20 million pieces, despite the fact that the company was founded relatively recently.

As of December 31, 2017, ZTO had more than 300,000 employees, 29,000 service outlets, 82 transit centers, and more than 950 network partners, with the volume of vehicles in more than 2000 trunk transportation lines exceeding 4800. Furthermore, the rate of accessible districts and counties in China was greater than 97.97%, and the coverage rate of outlets in townships was higher than 85.24%.

3.4.5.2 Network Distribution

As of December 31, 2017, ZTO Express had become a prominent, rapidly-growing private express delivery enterprise in China. The distribution of ZTO Express online stores is illustrated in Table 3.8.

Table 3.8 Distribution of the number of ZTO Express outlets

Province	Quantity	Province	Quantity	Province	Quantity
Guangdong	410	Liaoning	201	Xinjiang	126
Zhejiang	374	Neimenggu	184	Gansu	120
Hebei	356	Fujian	179	Tianjin	110
Yunnan	350	Shanghai	151	Guangxi	97
Heilongjiang	330	Anhui	150	Chongqing	91
Shanxi	313	Guizhou	149	Qinghai	54
Shaanxi	254	Jiangxi	147	Hainan	49
Sichuan	233	Jilin	146	Ningxia	39
Shandong	225	Hunan	139	Xizang	34
Henan	219	Hubei	138	Taiwan	1
Beijing	202	Jiangsu	136		

Source https://www.zto.com/express/expressWebsite.html

Figure 3.14 illustrates the distribution of ZTO outlets in various provinces and regions in China, as well as the corresponding number of outlets. In the Central China region, there are 496 outlets; in Southern China region, 556, in the Eastern China region, 1362; in the Northeastern China region, 677; in the Northern China region, 1165; in the Southwestern China region, 857; in the Northwestern China region, 593; and one outlet in the Hong Kong, Macao and Taiwan regions of China.

The Pearl River Delta is located in the central and southern parts of Guangdong Province and the mouth of the Pearl River. It is a major participant in the country's economic globalization, boasting rapid technological development and convenient transportation. As a result, ZTO Express outlets are densely distributed in the Pearl River Delta. The Yangtze River Delta, with Shanghai as the central city and Nanjing and Hangzhou as sub-central cities, is a hub of economic prosperity, with numerous transportation hubs. ZTO Express, established in Shanghai in 2005, has grown rapidly since then, with its outlets now densely distributed in various regions such as Zhejiang Province, Shanghai, and Beijing.

3.4.6 *Yunda Express*

3.4.6.1 Development of Yunda Express

In August 1999, Yunda Express, a renowned private enterprise in China, was founded. Adhering to the tenet of "creating value for customers, contributing to the society, and providing jobs for employees," Yunda Express has consistently innovated and expanded its business scope to 31 provinces (Districts and cities) in China, setting up over 70 transshipment centers and more than 40,000 service outlets.

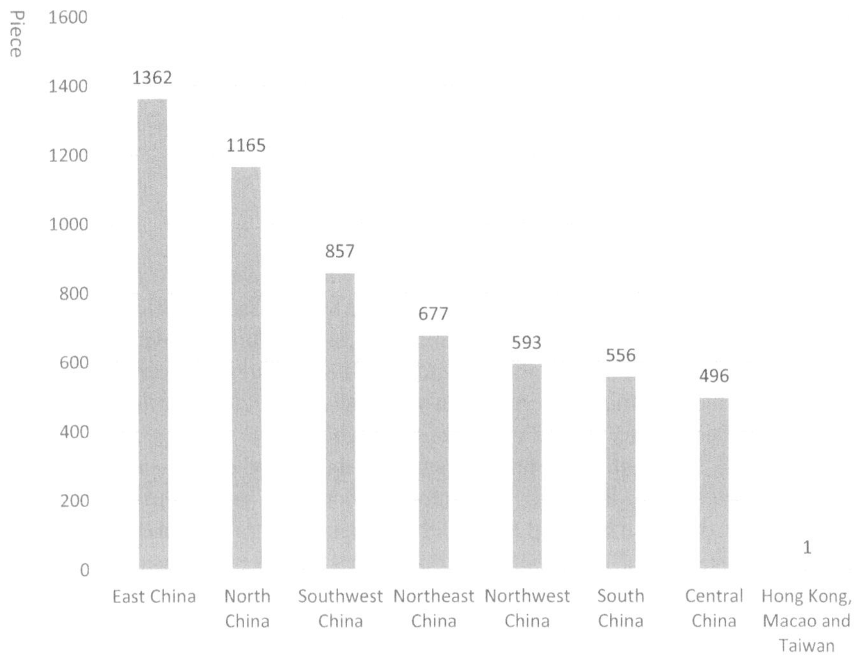

Fig. 3.14 The Regional Division of ZTO Express. *Source* The Author

Yunda Express has implemented a comprehensive strategy to improve the management of express delivery in every step, allowing customers to track package progress in real-time, as well as introducing the "service commitment" system for each link of the express collection, transfer, and delivery process. Furthermore, the company has employed technological means to supervise by remote video surveillance, etc. In 2013, a SMS reminder service was launched to make parcel sending and receiving more convenient for consumers. Additionally, stores have been constructed across China for customers to send and receive parcels with ease.

On June 6th, 2015, Yunda and SF Express made a joint investment in Shenzhen Hive Box Technology Co., Ltd. On February 28th, 2016, Yunda Express announced strategic and capital cooperation with financial investment institutions such as Ping An Insurance (Group) company of China, Ltd., Oriental Fortune Capital, and V Fund.

In October 2017, Yunda officially released its financial report for the third quarter of 2017, which revealed that the company's operating revenue amounted to 2.529 billion yuan, representing a year-on-year increase of 35.89%. Additionally, the net profit attributable to shareholders of the listed company was 428 million yuan, a year-on-year increase of 34.02%, while the basic earnings per share was 0.35 yuan, a year-on-year decrease of 5.41%.

In March 2018, Yunda Express ranked 37th in the BrandZ™ 2018 Top 100 Most Valuable Chinese Brands list.

Yunda Express will bolster the concept of proactive service and offer customers with more efficient, personalized, and convenient express services.

3.4.6.2 Network Distribution

Yunda Express has established more than 70 distribution centers nationwide and laid approximately 2500 trunk roads, 600 branches, and 800 air transportation routes. The company has established over 40,000 outlets across the nation, providing safe and speedy services, thus becoming the mainstay of private express delivery companies. According to the information from Yunda's official website, the distribution of their outlets as of December 31, 2017 can be seen in Table 3.9.

Figure 3.15 shows that the number and distribution of outlets in each region have been determined according to economic conditions and population density. Consequently, outlets in the Pearl River Delta, Yangtze River Delta, and central cities are highly concentrated. Transportation convenience must be increased in regions with smaller population densities, and high costs of expenses in the western region have caused a low number of customers and outlets.

Table 3.9 Distribution of the number of Yunda Express outlets

Province	Quantity	Province	Quantity	Province	Quantity
Jiangsu	4926	Shanghai	1650	Yunnan	643
Guangdong	4121	Shaanxi	1495	Xinjiang	401
Zhejiang	3854	Heilongjiang	1443	Gansu	379
Fujian	2448	Sichuan	1317	Tianjin	331
Henan	2411	Jiangxi	1249	Neimenggu	279
Anhui	2378	Liaoning	1067	Hainan	246
Shandong	2005	Chongqing	1650	Ningxia	131
Beijing	2005	Hunan	1495	Qinghai	70
Hubei	1989	Guizhou	1443	Hong Kong	60
Shanxi	1976	Jilin	1317	Xizang	59
Hebei	1675	Guangxi	1249	Taiwan	3
				Macao	1

Source http://www.yundaex.com/cn/fuwuwangdian.php

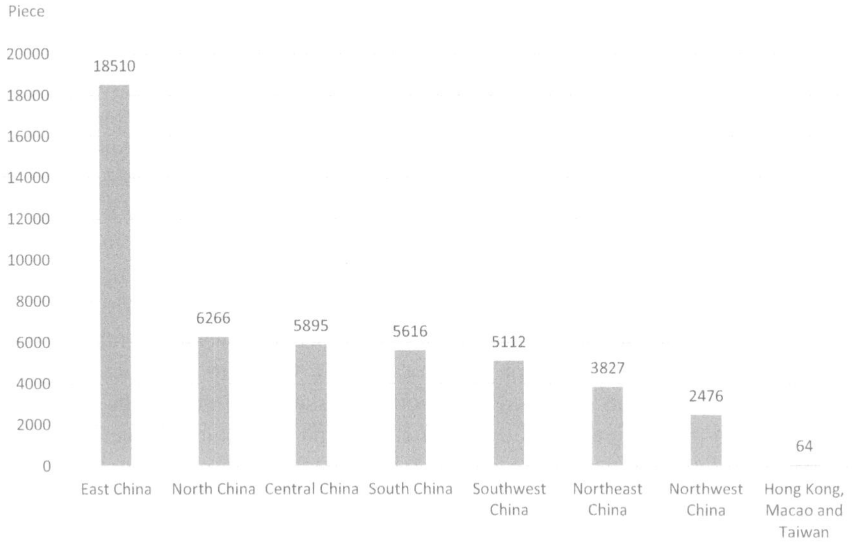

Fig. 3.15 The Regional Division of Yunda Express. *Source* http://www.yundaex.com/cn/fuwuwangdian.php

3.5 Implications of the Development of the Express Delivery Industry

3.5.1 Advantages of International Express Delivery Enterprises Development

Product innovation is the root of the rapid development of express delivery enterprises. Yamato Transportation kickstarted the rapid development of express delivery enterprises in 1983 by introducing ski equipment delivery services for consumers. Following this, in 1984, it innovated a new type of express delivery business, distinct from ordinary transport businesses, such as golf equipment. Other express delivery companies have since followed suit, introducing various special delivery services according to their own characteristics. For example, DHL provides basic express services for Deutsche Telekom AG as well as comprehensive value-added services, such as machining of mobile phone parts and handling of replacement parts for other products.

An oligopoly market is an important means for enterprises to expand and occupy the market. In the highly competitive U.S. express delivery industry, the current issues are addressed by acquiring or agreeing to acquire other express delivery enterprises, thus causing the market structure to shift from multi-enterprise competition to oligopoly competition. In Japan, the express delivery market was dominated by the three major express delivery enterprises, Yamato Transport, Sagawa Express, and Japan Post, who held a market share of 84.4% as of 2005, making the oligopoly market.

Investment in information technology is an important direction for express delivery companies. Express delivery enterprises have increasingly complex operations due to the involvement of various transportation methods, and customers have high expectations for the quality of express service. To meet these demands, Japan's Yamato Transportation has implemented large-scale automated sorting machinery in their cargo sorting centers to improve express efficiency. In Germany, the express delivery industry is becoming more intensive, digitized, and interconnected. Investing in informatization is therefore a key priority for express delivery enterprises.

It is important to establish a strategic partnership and build operating outlets. REA has collaborated with the telegraph operator Western Union to construct low-cost nationwide outlets. Japan's Yamato Transport has established more than 3 million basic outlets in places frequented by homemakers and 24-h convenience stores, to serve as centralized service outlets for goods before delivery to consumers.

3.5.2 *Problems in the Development of China's Express Delivery Industry*

Currently, there are many issues concerning the development of China's express delivery industry. Due to the booming economy in our country, the demand for express delivery services has increased, resulting in the rapid growth of numerous small and medium-sized private express delivery enterprises. However, a complete system has not been established yet, and the express delivery market is still fragmented. The lack of prominent leading enterprises and the emergence of a "multi-blend" mode are largely due to the fact that most enterprises are adopting low-cost competition strategies (Wang, 2015).

From a resource investment perspective, China's express delivery enterprises have only increased the use of cheap labor to meet their business expansion needs, and have not devoted resources to researching new scientific management modes and emerging information technologies to reduce operating costs. As a result, many express delivery enterprises still require further refinement in their employment system, resulting in increased staff mobility and increased training costs for professional skills. At the managerial level of the enterprise, the shortage of high-end talents has resulted in a lack of capacity for the development of knowledge-based express delivery enterprises. The failure to create efficient and effective professional skills training procedures directly affects the service quality of express delivery, leading to issues such as delays, damage and loss which can have a severe impact on express delivery enterprises' brand influence and hinder their competitiveness and development in the market.

From an enterprise growth point of view, many Chinese express delivery companies have adopted the model of collaboration. This resulted in a low entry threshold to the express delivery market, and has increased the number of Chinese express delivery companies dramatically. But there are a maximum of 25 having any significant brand recognition. This chapter examines nine of the most popular express delivery companies, and it was found that they still have a disadvantage in the international market due to their careless expansion, causing a drop in service quality and a decrease in brand competitiveness. Due to the fact that they are unable to guarantee service quality, they can only expand their market share through adjusting prices, resulting in a decrease in the profits, further drops in service quality, and an increase of problems within the industry.

The development gap between China's international express delivery business and that of global express delivery companies is very large. As shown in Table 3.10, the overall strength is not strong, particularly the air transportation capacity is much lower than that of international express delivery companies. This has caused Chinese express delivery businesses to be less competitive in the international market. Cases in point are China Postal Express and Logistics, a state-owned enterprise, and SF Express, a fast-growing private company.

Based on the development history and advantages of international express delivery enterprises, as well as the existing problems in the development history of the express delivery industry in China, the following insights can be gained.

(1) Advanced information technology is to be actively introduced. SF Express, a representative of private express delivery companies, has introduced advanced information technology to offer customers a national unified service hotline for placing orders and consulting directly. This technology provides enterprises with safe express services, reduces operating costs, improves customer satisfaction, and enhances the future competitiveness of express delivery companies.
(2) The construction of distribution centers is to be strengthened. In 2015, major express delivery companies strengthened the construction of distribution centers, with the aim of optimizing the operation process and increasing the proportion of automatic and semi-automatic sorting equipment to improve express processing capacity. Such distribution centers as Postal EMS Central China Mail Processing Center, YTO Express Northwest Transshipment Center, and JD Guiyang "Asia No. 1" Logistics Center were built and expanded in number, then officially put into operation. The operation process was to be optimized. The proportion of automatic and semi-automatic sorting equipment were added, and the express processing capacity was improved then.
(3) Terminal delivery methods are to be innovated. Express delivery companies have increased their efforts to explore terminal distribution models. SF Express, YTO Express, and other companies have expanded their terminal distribution networks with services such as "SF Home," "Mama Station," and "Express + Convenience Store." YTO Express and ZTO Express have made efforts to standardize their urban outlets, improve service standards, and deepen their strategic cooperation. Express delivery companies are strengthening their terminal service capabilities through collaborative construction. SF Express, STO Express, ZTO Express, and other companies have jointly established

Table 3.10 Comparison of delivery capabilities of China's and international express delivery companies

	Headquarters	Service area	Transport capacity	Operating facilities
UPS	U.S.A., Atlanta	More than 220 countries and regions and cover North America and Europe	237 self-owned aircraft, 41 leased aircraft and 106,024 transport vehicles	1990 operating facilities worldwide, 15 air freight centers
FedEx	U.S.A., Memphis	More than 220 countries and regions around the world	More than 250 aircraft, 47,000 transport vehicles	1200 service stations around the world, 12 transfer centers
DHL	Germany, Bonn	More than 220 countries and regions around the world	More than 250 aircraft, 80,000 transport vehicles	More than 3000 service centers and 35 international transit centers
EMS	China, Beijing	More than 200 countries and regions around the world, cities within 31 China's provinces	9 only-cargo aircraft, more than 20,000 delivery vehicles	More than 45,000 business outlets, and a large distribution center in Nanjing
SF Express	China, Shenzhen	Chinese Mainland and Hong Kong, Macau and Taiwan, the United States, Japan, South Korea, Singapore, Malaysia, Thailand, Vietnam, Australia	15 self-owned cargo planes, more than 120,000 transport vehicles	More than 9100 business outlets in the world

Source The official website of each company

Shenzhen Fengchao Technology Co., Ltd. Additionally, some express delivery companies are using third-party outlets to extend their terminal networks in order to enhance their terminal service capabilities.

(4) The external environment has a positive impact on the express delivery business. As competition in the market intensifies, the demand for capital in the industry has also increased. Social capital has been injected into the express delivery industry through equity investment in express delivery companies. For example, Alibaba has partnered with Yunfeng Fund to invest in YTO, while Warburg Pincus, Hillhouse Capital, Standard Chartered and Sequoia Capital have all invested in ZTO Express. The influx of diverse capital has provided financial support in allowing the industry to expand. As consumer needs and concepts evolve, the scale and transaction volume of cross-border e-commerce have grown significantly, bringing new opportunities to the express delivery industry. Both cross-border e-commerce and express delivery are intertwined, forming an ecosystem which mutually affects and restricts each other.

(5) The manufacturing industry is essential for the growth of express delivery services. In turn, express delivery services are a key factor in the transformation and improvement of the manufacturing industry. By working with the manufacturing industry, express delivery companies are able to integrate functions and extend their reach, allowing them to transition from express delivery operators to comprehensive express logistics operators, thus helping manufacturing enterprises reduce costs, increase efficiency, restructure, and transition from large to solid enterprises. The combination of the two industries is progressing at a quickened pace. Companies like EMS and SF Express have established long-term and secure partnerships with large manufacturing companies. Express delivery companies are also actively settling in industrial parks, providing close-range, all-round, and integrated services to manufacturing enterprises.

3.6 Summary

This chapter primarily focuses on the progress of the express delivery industry, as well as the progress and current state of global express delivery companies and China's state-owned and private enterprises. It provides an analysis of several well-developed international express delivery companies, and summarizes their progress and current state in China. It also outlines China's express delivery market, the development level of major express delivery companies, and the state of express delivery network construction. Lastly, it provides a comprehensive analysis of international express delivery companies and China's current state. Additionally, this chapter investigates the issues and current advantages of China's express delivery companies in the development process.

References

Fan, C. H. (2007). *Research on CRE logistics network planning based on Hub and Spoke Theory* [Master dissertation, Beijing Jiaotong University].

Guo, H. (2013). *Research on development strategy of express delivery business of China postal express logistics company* [Master dissertation, Beijing Jiaotong University].

Li, Z. H. (2015). Research on the selection of SF Express operation mode. *Logistics Engineering and Management, 37*(7), 47–50.

Lu, S. (2010). Inspiration from the development of German express industry. *China Logistics and Purchasing, 2*, 6–7.

Wang, F. F. (2015). *A comparative study on the development mode of domestic and international express delivery industry* [Master dissertation, Hebei University of Technology].

Chapter 4
Express Delivery Network Structure and Characteristics

Express delivery network is a unified whole, and it is the basis for the operation and development of express delivery companies to have a well-organized, scientifically managed, efficiently operated, and coordinated-controlled network structure. The choice of the network structure model and its corresponding characteristics are of strategic importance in optimizing express delivery network. This chapter initially gives an in-depth explanation of the composition and topology of express delivery network, then evaluates the characteristics of express delivery network structure and express delivery traffic flow, laying the groundwork for the following chapters.

4.1 Express Delivery Network Structure

Express delivery network is a unique form of transportation system which operates according to specific rules for organizing regional transit centers, city transit depots, service outlets and all routes of the network under unified management and scheduling. The goal is to deliver goods to customers (Zhang & Nie, 2009). A network has nodes that represent cities, and network edges represent transportation trunks between them. express delivery network topologies consist of nodes and transportation routes, which come in two forms: Fully-connected (FC) express delivery network and hub and spoke (HS) express delivery network (Bai & Zhu, 2006).

4.1.1 Composition of Express Delivery Networks

An express delivery network usually consists of five components: Regional Transit Centers, City Transit Depots, Service Outlets, Transportation Routes (Networks), and Customers (shown in Fig. 4.1) (Yang et al., 1996). Regional Transit Centers serve as

P. Li and J. Wu, *Research on Innovation of Express Delivery Network Management*,
https://doi.org/10.1007/978-981-99-8982-9_4

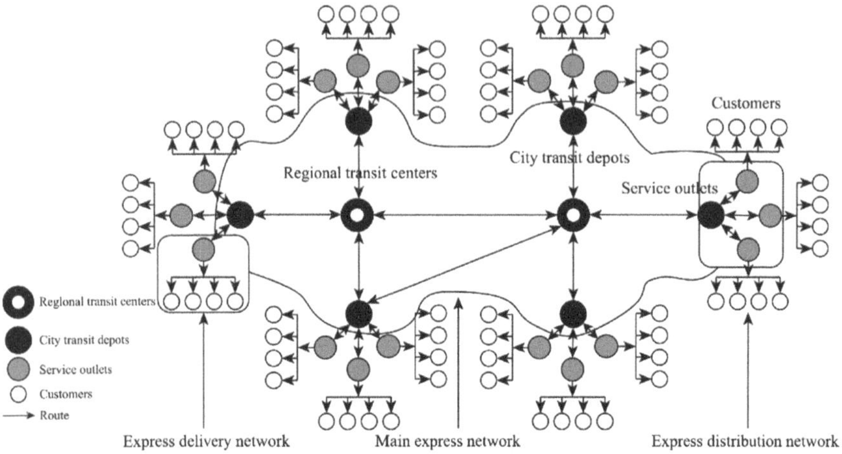

Fig. 4.1 Overall structure of express delivery network. *Source* The Author

the "brain" of a network, City Transit Depots as "nerve hubs", trunk lines between provincial capital-level cities as "aorta", and lines between prefecture-level cities as "blood vessels", forming the network like capillaries at a lower level. In actual operation, express enterprises add terminal outlets, such as self-collection outlets, community service stations and self-collection cabinets (such as Cainiao Station, Postal outlets, HIVE BOX, etc.). So network nodes from regional transit centers to terminal outlets can be divided into four levels (Lin & Xie, 2017).

4.1.1.1 Regional Transit Center

A regional transit center is an essential part of express delivery network, as it enables the sorting, transferring, and distributing of goods. More specifically, it can process, sort, and transfer inter-provincial and local packages to other outlets or urban transit depots. This helps to facilitate fast transportation and efficient connections of different forms of transport. These centers, also known as regional sorting and distribution centers, are typically located near transportation hubs of inter-provincial highways. When selecting the sites for regional transit centers, builders usually take into account the socio-economic environment, traffic conditions, demand for goods, and connections with other regional express nodes.

4.1.1.2 City Transit Depot

The City Transit Depot, also referred to as the City Sorting Center, City Transfer Center, and City Distribution Center, is a distribution hub in the express delivery network. It is mainly responsible for collecting, sorting, dispatching, and distributing express packages in a specific urban area to subsidiary express service outlets. At the same time, it also has a service network that delivers goods directly to customers in the surrounding areas. Additionally, express enterprises have high requirements for their transit centers with regard to mechanization, intelligence, and information level. The size and efficiency of transit centers vary by region. They are divided into three categories based on priority: primary, secondary, and tertiary. There are also major variations in the service offerings and express package handling capabilities between the different levels of city transit depots. The primary-level transit station has a wider range of services and can process more efficiently. It is responsible for several secondary-level transit stations as well as other connected outlets. Therefore, the hierarchy of express delivery network transportation depends on the density of city transit depots. Additionally, when deciding on a location, factors such as traffic conditions, freight flow, operational efficiency, marginal cost, urban planning, local policies, and connections with other city transit depots within the coverage of the higher-level nodes have to be taken into account.

4.1.1.3 Service Outlets

Service outlets focus on providing pick-up and delivery services, located at the beginning and end of an express delivery network. These outlets are distributed within the service range of express delivery companies to obtain customers information and needs in real time. Each outlet must collect items from customers and deliver them to recipients. When deciding the location of these outlets, traffic conditions, geographic information, population distribution, and various costs of the candidate region should be taken into account.

4.1.1.4 Transportation Route

During express transport, the route taken by the vehicles carrying packages between nodes (pick-up and delivery outlets, city transit depots, and regional transit centers) is known as the transportation route. The most common means of transportation are railway, road, and air. The type of transportation used for different routes will vary. The transportation route connects various points in the express delivery network, creating a full express delivery network. The transportation route is split into two layers based on the connecting nodes: the first-level network and the secondary network. The first-level network consists of transportation routes between transit centers of the same level, such as ZTO Express which has created more than 60 regional transit centers in China. Large trucks are usually used for road or air transportation between these centers. The secondary network involves transportation routes between city transit depots, with rail or car transportation used on city expressways or high speed. The lines connecting the city transit depots and the service outlets are called branch lines, which are typically serviced by minivans or other reliable transportation.

4.1.1.5 Customers

Customers are the first people to use the pickup network and the last to use the delivery network. With fierce competition in the present era, customers increasingly expect more from express services. To improve the efficiency of express delivery networks, express enterprises must have more precise management procedures and reduce transportation time. Customers are at the heart of the express delivery network, so effective operation and management of the network should focus on effectively collecting packages, tracking and analyzing related information, meeting the needs of each customer, communicating with them appropriately and increasing their satisfaction with express services in the shortest time possible.

4.1.2 Express Delivery Network Topology

4.1.2.1 Common Network Topology

Network topology is a way of expressing the elements in a network and the connections between them. We examine the manner, method, and shape of the interconnection of each endpoint in the network system structure, which is based on the two basic elements of "point" and "line" from geometry. Figure 4.2 illustrates that networks are divided into bus, star, ring, tree, and mesh topology, depending on the different topological structures. All modern networks have evolved from these basic networks. Table 4.1 outlines the advantages and disadvantages of network topology.

4.1.2.2 Topology of Express Delivery Networks

The express delivery network topology has evolved from the primary network topology. It is not feasible to investigate all topologies of the express delivery networks due to the large number of Chinese express enterprises and their various service standards. This investigation is mainly based on a large Chinese express enterprise that has a comprehensive express service and network system, which is similar to the operation management of other Chinese express delivery companies and can be used as a reference.

(1) Collection of Express Data

Accurate data is needed in abundance to analyze the express delivery network's spatiotemporal dynamics and complex structure. Table 4.2 provides the information of an express enterprise, which includes the time and place of express pick-up, sorting, distribution, loading, unloading and delivery. After data screening, cleaning and sorting, the distribution of logistics nodes (regional transfer centers/city transit depots) and the connectivity between enterprise nodes across China are obtained. The

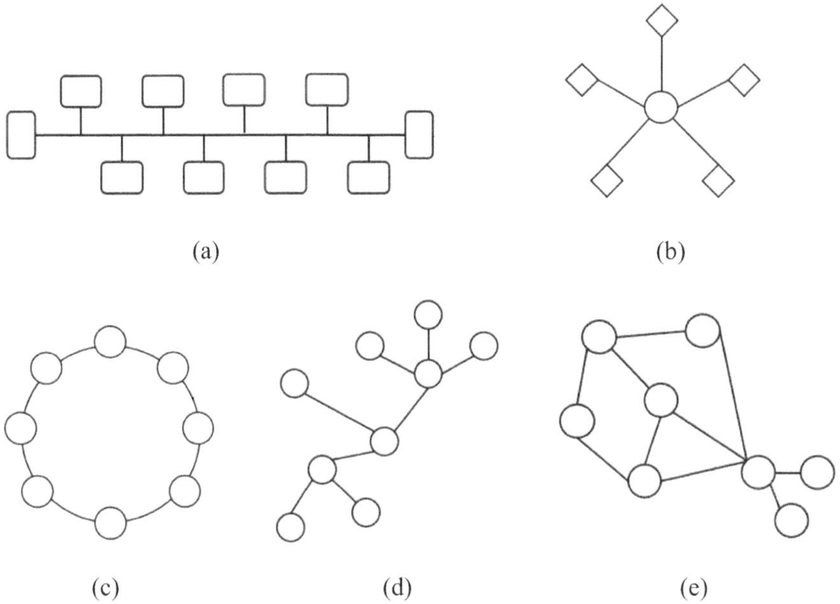

Fig. 4.2 Common network topologies. **a** Bus topology. **b** Star topology. **c** Ring topology. **d** Tree topology. **e** Mesh topology. *Source* The Author

data is distributed by the hour in time and by cities in space, providing a comprehensive view of the express delivery network's spatiotemporal dynamics in transportation and allowing for the analysis of its topological features.

The following explanations must be made about the obtained data: (1) The distribution network can be thought of as an undirected network, meaning that regardless of which direction the delivery is going, from node A to B or node B to A, there is only one delivery route between the two; (2) It can be considered as an unweighted network, meaning that the quantity of distribution between cities is not taken into account or the weight of the edge is not taken into consideration; (3) No matter what mode of transportation is used between cities, we only consider if there is a connection between them, or if there is a distribution route; (4) Factors such as the time, cost of delivery, and the operational efficiency of express enterprises do not need to be considered.

(2) Construction and Analysis of Express Delivery Network Topology

We analyze the topological characteristics of the express delivery network by filtering and integrating express data, summarizing the topology connection between all nodes, and constructing the network's topological structure. Additionally, we examine how different topological characteristics impact express delivery traffic flow.

First, a preliminary screening of the data is conducted. If the volume of express between two nodes exceeds the standard threshold, then an "edge" is determined to

Table 4.1 Advantages and disadvantages of common network topologies

Network topology name	Advantage	Disadvantage
Bus topology	It has a simple structure, which is easy to construct, and its overall performance is quite remarkable. Any node can be bidirectionally transported	Bus line failure is easy to cause the paralysis of the overall network structure
Star topology	It has flexible structure, which is easy to build and manage. The surrounding nodes are controlled by the central node. When a node in the structure fails, it is easy to detect and the fails will not have a great impact on the overall structure	Central node withstands heavy burden. Usage of route is poor
Ring topology	It has simple structure. Easy to manage the established network. All nodes locate at the same level	Direction is single. It has too many nodes, which affects transportation efficiency
Tree topology	Nodes support two-way transportation. This topology has flexible transportation and low cost, It is suitable for hierarchical management systems with primary or secondary levels	It is difficult to centralized control, if there is a fail, it is hard to find
Mesh topology	The structure is complex and diverse, the node connection is arbitrary and disordered, and the connection route between each nodes not only one. It is suitable for systems with large business volume	Construction cost is high, structure is complex, and it is difficult to adopt centralized control

Table 4.2 Express logistics information format

Date	Time	Shipment status	Current location
2018-10-17	17:23:39	[Xi'an Chang'an Road] Package has been collected	Xi'an
2018-10-17	18:41:39	Packages left [Xi'an Chang'an Road] and sent to [Xi'an]	Xi'an
2018-10-17	19:57:24	Packages has arrived at [Xi'an Transit]	Xi'an
2018-10-17	20:09:54	Package left [Xi'an Transit] and sent to [Dongwan Centre]	Dongguan
2018-10-19	03:17:42	Package has arrived at [Dongguan Centre]	Dongguan
2018-10-19	03:53:19	Packages left [Dongguan Center] and sent to [Huidong]	Huidong County
2018-10-19	09:26:01	Package has arrived at [Huidong]	Huidong County
2018-10-19	10:09:50	Sending package…	Huidong County
2018-10-19	13:25:36	Package has been signed for in [Huidong], the recipient: Anonymous	Huidong County

Source The Author

exist between them, and the size of the express volume determines the weight of the edge. Secondly, there is bi-directionality between nodes; packages can travel from node A to node B and vice versa. At this point, the express delivery network G can be regarded as a weighted, bidirectional network, denoted as G = (V, E). V represents the node set, and E represents the edge set, with at least one edge in E corresponding to each node in V.

Figure 4.3 shows the express delivery network G created by the ForceAtlas2 layout of Gephi software. The nodes in the figure represent the prefecture-level cities involved in the express data, while the edges indicate the actual express delivery routes. The size of each node reflects its degree, with the high-degree nodes more concentrated in the center of the network, and the low-degree nodes at the edge. The nodes in the core layer are almost exclusively connected structures, primarily regional transit centers, which form the core of the express delivery network structure. Most of those located in the backbone layer are secondary and tertiary-level city transit depots. They are partially connected with the nodes in the core layer (usually the provincial capital city), and there are few nodes outside the backbone layer. The bottleneck node is a transition node, meaning that some nodes can only communicate with the outside world through it. For instance, Lanzhou, the provincial capital of Gansu Province, has various transportation options for getting out of the province. On the other hand, prefecture-level cities such as Dunhuang and Liangzhou rely on the bottleneck node to deliver express. Therefore, if there are any issues at the bottleneck node, it will negatively impact the delivery of express in Gansu Province.

This express delivery network topology is a hybrid combination of star, tree, and mesh topologies. Tree topology is often used for regional delivery, for example, for

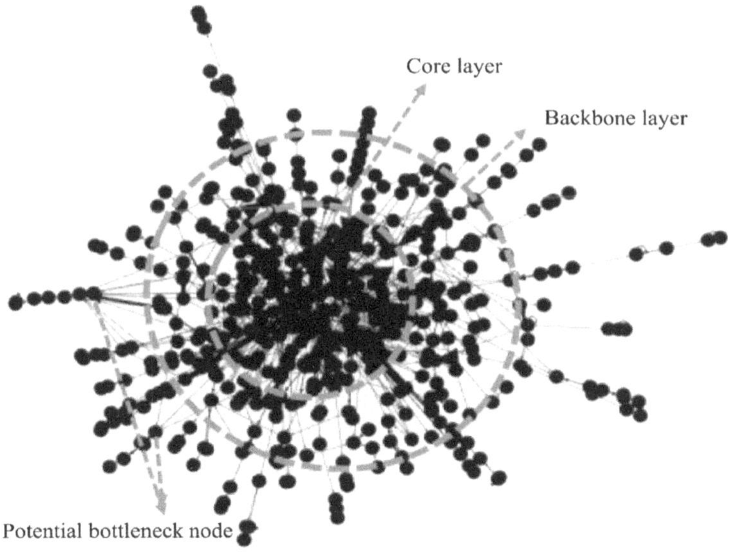

Fig. 4.3 Express delivery network topology. *Source* The Author

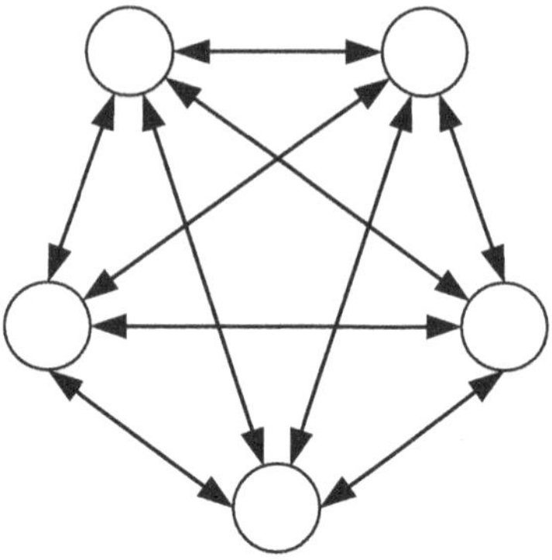

Fig. 4.4 Fully-connected express delivery network. *Source* The Author

transportation from regional transit centers to city transit depots, service outlets, and customers. This hierarchical structure is beneficial for distributing responsibilities. A star topology is often used between hierarchical nodes, which helps enable centralized control, for example, between city transit depots and various service outlets. Furthermore, each node to the terminal customer node often uses a mesh topology, which is dynamic and random. Additionally, optimizing the mesh topology is essential for reducing costs, shortening transportation time, and improving the efficiency of express delivery network operations.

Figure 4.3 shows two typical structures of express delivery networks: the hub and spoke express delivery network and the Fully-connected express delivery network.

(3) Fully-connected Express Delivery Network Structure

Based on the express delivery network topology, the network structure has been further studied. Figure 4.4 shows the structure of a fully-connected express delivery network, where the "nodes" represent the cities that the express service covers and the lines between them represent the transportation routes between the two cities.

A fully-connected network is a complete graph network in which packages are able to be sent directly from any two nodes without any transfer. This allows for a direct path between any two cities, eliminating the need for a transfer process and thus improving the efficiency of distribution. Additionally, this network is both random and arbitrary. The limitations of a Fully-connected Network are the number of outlets available and the amount of transportation it can handle. When there are a lot of service outlets, transportation can be done quickly without having to plan the routes, but the cost of transportation will go up and the full-loaded rate won't be guaranteed. This makes it hard to manage transportation routes, like in the case of air express. So, a fully-connected network is only suitable for express delivery networks that have a few service outlets and large amounts of transportation.

(4) Hub and Spoke Express Delivery Network Structure

The hub and spoke express delivery network is a transit transportation structure in which express items are not sent directly to their destinations (Li, 2006). Instead, they are transported from their origin to a hub node, and then to their destination with at least one transit in between. The hub nodes play an important role in this network structure, as they not only facilitate transmission between different nodes, reducing the number of transportation routes, but also take some of the pressure off core nodes (such as national distribution centers), making the entire network more stable.

The topology of a Hub and spoke express delivery network depends on the connections between nodes. This type of network has two standard assumptions: non-hub nodes are not connected to each other and all hub nodes are connected. Hub and spoke express delivery networks can be divided into single hub and spoke networks and hub and spoke networks with multiple hubs, depending on the number of hub nodes. Single hub and spoke networks are typically found in prefecture-level cities with poor transportation, such as Xizang and Xinjiang. On the other hand, multiple hubs are the current development trend of express delivery networks. As shown in Fig. 4.5, large and medium-sized cities have established transit hubs that connect transportation lines. These transit centers are directly connected to their closest hub centers, allowing for express to be transmitted directly to various locations. The hub and spoke express delivery network can compensate for the shortcomings of fully-connected networks and address issues such as excessive transportation lines, lack of guaranteed full-load rate, and wasted transportation resources. Unified operation through transit hubs has improved transfer capacity and efficiency. However, the entire express delivery network has extremely high expectations for the transit hub's capability and effectiveness in delivering packages. Therefore, transit hubs could potentially become a constraint in the express delivery network structure.

The hub and spoke network that includes multiple hubs can be divided into single-distribution and multi-distribution, based on the relationship between each node and the hub. As shown in Fig. 4.6, in single-distribution hub and spoke network, each node is connected to only one transit hub, whereas in multi-distribution hub and spoke network, each node is connected to at least one transit hub.

In the 1880s, FedEx first attempted to implement the hub and spoke express delivery network, which increased the utilization rate of transport resources and reduced transportation times and costs (Chen et al., 2010). This successful application of the hub and spoke express delivery network led to other express enterprises trying to imitate this model, resulting in hub and spoke express delivery network becoming the primary form of transportation in the express delivery network. As the express transportation mode has matured, the express delivery network has transitioned from a Fully-connected express delivery network or hub and spoke express delivery network structure to a hybrid topology combining both structures (Fig. 4.7). The hub and node are fully connected and common nodes can be connected to each other or directly connected to the transit hub (Ni, 2010).

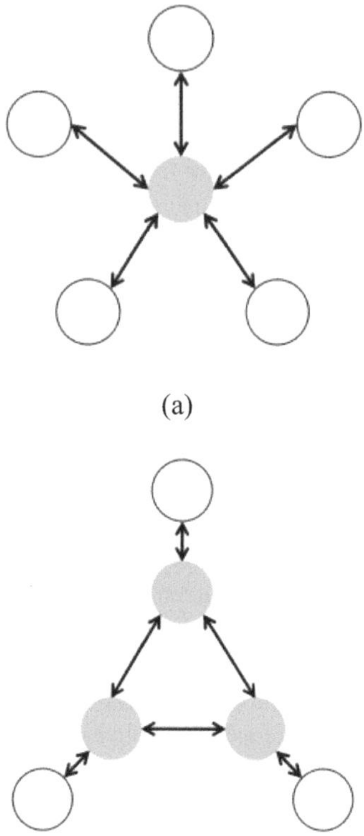

(a)

(b)

Fig. 4.5 Hub-and-spoke network classified by number of hubs. **a** Hub-and-spoke express delivery network with single hub. **b** Hub-and-spoke express delivery network with multiple hubs. *Source* The Author

As an example, Fig. 4.8 displays the express delivery network of a major express company in Shaanxi Province. The primary central node of the network is located in Xi'an, while prefecture-level cities are the secondary nodes. Terminal nodes are located in counties, while the majority of townships do not contain nodes. The network has a hybrid hub and spoke structure. The hub and spoke as well as the fully-connected express delivery network have developed from a star topology. The edges between the different star topology networks are connected, creating a hub node, and this allows for a fully-connected express delivery network to be formed in order to meet the delivery requirements. In a star topology network, lines are created between regular nodes and hub nodes to meet the delivery requirements, forming a fully connected express delivery network.

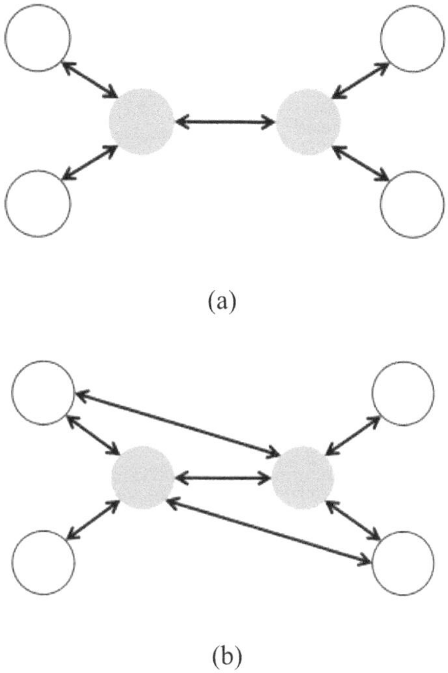

Fig. 4.6 Hub-and-spoke express delivery network classified by distribution relationship. **a** Single-distribution hub-and-spoke express delivery network. **b** Multi-distribution hub-and-spoke express delivery network. *Source* The Author

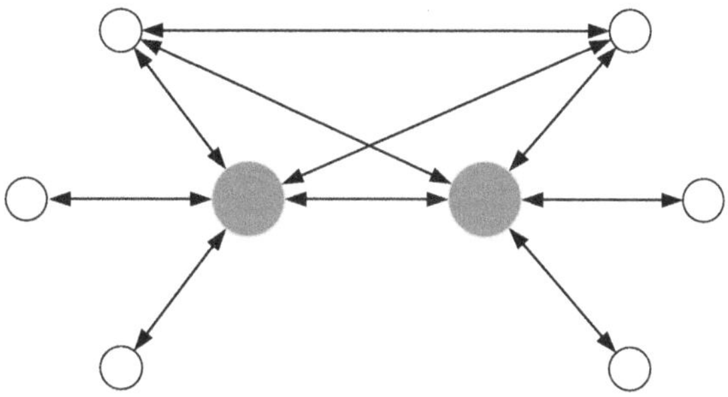

Fig. 4.7 Fully connected—hub and spoke hybrid express delivery network. *Source* The Author

4.2 Structural Characteristics of Express Delivery Networks

4.2.1 Characteristics of Backbone Networks

The backbone network of an urban express transportation system is essential, as it is responsible for "passing" traffic while providing service functions. This means that there must be a high number of express vehicles that are fast. The backbone network has the following characteristics.

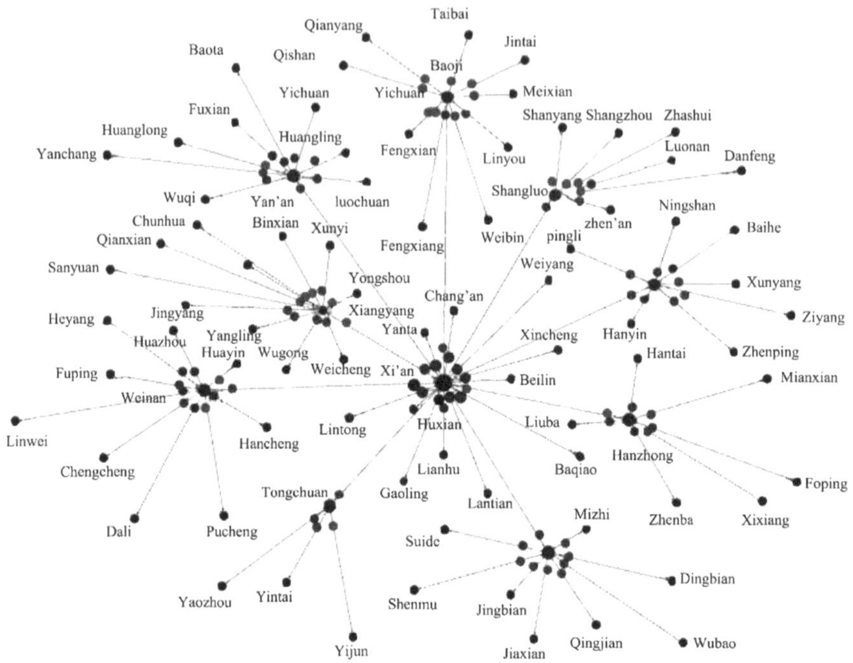

Fig. 4.8 The express delivery network of an express company in Shaanxi Province. *Source* The Author

4.2.1.1 Bidirectionality

Bidirectionality allows packages to be delivered in either direction between node A and node B, with multiple routes available. This maximizes utilization of the express delivery network, and thus helps to conserve resources of the main network.

4.2.1.2 Sharingness

The transportation network has developed rapidly as a whole, meeting the needs of people in various regions for convenient travel. However, on a deeper level, the degree of sharing between the two structures is quite different, despite both being bidirectional.

Under the fully-connected express delivery network structure, the sharing degree of each transportation route is not high, with each route usually connecting only two nodes. The number of transportation routes in the entire transport network is not less than the number of nodes, meaning that any node under this structure can be accessed directly without transit. Additionally, the failure of any side in the network structure does not affect the connectivity of the whole network, which has the advantages of short distribution time, high reliability, and low delay.

Under a hub-and-spoke express delivery network structure, there is a high degree of sharing among transportation routes. This is mainly manifested in three ways: first, multiple nodes share a transportation route with a high utilization rate; second, expanding transport branches to create a larger structure is made easier; and third, it is easy to isolate faults. If a particular branch fails, it can be isolated without affecting the entire system.

4.2.1.3 Inequality of Resources

The network distribution is seriously uneven on a deeper level.

On the one hand, under a fully-connected express delivery network structure, the density of the traffic network is small and the transportation routes between hub nodes are easily visible. However, the traffic network under a hub and spoke express delivery network structure is denser and broader. As illustrated in Fig. 4.9, the relatively large nodes are hub nodes, which are usually connected in a fully-connected express delivery network structure, providing the strongest accessibility between the nodes. The hub and spoke express delivery network primarily connects ordinary nodes to hub nodes, and ordinary nodes are unable to be connected to each other directly.

Table 4.3 shows that the infrastructure resources of road and railway networks in different regions are unequally distributed, with Eastern China having the highest density, followed by Central China, the National average, and Western China having the lowest density. Specifically, the road network density in the East is more than six times the West's, 1.9 times the Central China's, and 2.9 times the National average. Moreover, the railway network density in the East is 4.3 times the West's, 1.5 times the Central China's, and almost 2.5 times the National average.

In conclusion, the level of development and basic service capabilities of transportation networks in various regions of China vary greatly, with the eastern region having a clear advantage in terms of transport service capabilities and network coverage compared to the central and western regions. However, the density of the road network in the western China region is only 57% of the national average, and 19% of the eastern region; the density of the railway network is also 57% of the national average and 23% of the eastern region. Consequently, the development level of the western China region is far behind the national average and the development levels of the central China region, and it is necessary to further develop infrastructure according to the actual situation.

4.2.1.4 Economies of Scale

The concentration of network traffic can reflect economies of scale. Economies of scale are reflected when a given technology is adopted regardless of its product specifications, and the average cost is reduced within the output range. This can be

Fig. 4.9 The express company's topology. *Source* The Author

Table 4.3 Statistics of national road network density

Region	Highway mileage (km)	Railway mileage (km)	Density of highway network (km/100 km^2)	Density of railway network (km/10,000 km^2)
Total of China	4,773,469.00	126,970.00	49.72	132.20
Eastern China	1,555,323.00	39,524.00	146.50	327.28
Central China	1,273,951.00	35,546.00	76.29	212.86
Western China	1,944,222.00	51,900.00	28.26	75.45

Source The Author

considered an economy of scale, while the opposite is uneconomical. Similarly, there is a difference in the transport capacity between nodes.

For a hub-and-spoke network, the express is usually collected at the hub and then transported to its destination through the hub, taking advantage of the transport capacity of that road section to give the main line in the network a scale effect of economic. This also reduces the investment, operation, and maintenance of facilities and equipment in other road sections. Additionally, larger vehicles are used to transport the express collected from different directions together, increasing the full-load rate and reducing the unit transportation cost.

Compared to the hub and spoke network structure, the economies of scale for the fully-connected network structure needs to be improved. In practice, the demand for single routes is low and rarely reaches its full-load capacity, leading to a higher transportation cost and the wasting of resources. Furthermore, the expansion of branches within the fully-connected network structure makes it difficult to open up new transportation routes that are shorter and less costly, which prevents the express traffic volume from increasing.

For example, Fig. 4.4 shows a fully-connected network with 5 nodes and 10 transportation routes, requiring at least one vehicle for its entire transportation route. In contrast, Fig. 4.5a shows a hub and spoke network with only 5 transportation routes and a relatively short entire transport route, requiring at least 5 transportation vehicles.

4.2.2 *Node Features*

The positioning, number, service level, and other essential requirements of nodes in the express delivery network structure are essential to efficiently undertake and complete the regional express business and meet consumers' demands for timely delivery. Given that the infrastructure of an express delivery network is not perfect at present, nodes' affordability, reliability, and control ability are direct indicators of the express delivery network structure.

4.2.2.1 Different Affordability of Nodes

In the fully-connected express delivery network structure, there is no clear distinction between primary and secondary nodes, as most are hub nodes and the number is relatively small. This means that the cost of each node is similar and the burden is minimal. However, in the hub and spoke express delivery network structure, common and hub nodes are clearly distinguished, allowing resources to be allocated to each node in a rational way according to different levels and making the most of them. Unfortunately, each single node is highly loaded and carries a heavy burden, and any failure will affect the whole network.

For example, the hub nodes in a hub and spoke network structure could be regional transit centers and city transit depots. A regional transit center is a comprehensive information processing center, such as Guangzhou in South China, which radiates out to Hainan, Guangxi and Guangdong Province. It is responsible for outlet management, cargo storage, declaration, sorting, packaging and distribution within the region. The city transit depots, such as Humen and Dongguan, are nodes with developed regional transportation and relatively concentrated express enterprises, with abundant express routes and convenient cargo transit. These hub nodes are responsible for the express volume of the entire region. If express packages are hoarded or transit hubs become overloaded, the overall operating capacity of the express delivery network will be reduced.

The service outlets which are the ordinary nodes of a hub and spoke network structure have capacity that is demonstrated in the following two ways.

(1) The number of people being served by an outlet is decreasing.

Figure 4.10 illustrates that the population density of national express service outlets has decreased from 2012 to 2017. In 2012, the average amount of express service outlets per 100,000 people in China was 6.9, and by 2017, this number had increased to 16. This suggests that the number of people being served by each unit has decreased, thereby enhancing the per capita service capacity. The downward trend has yet to reach a stable state, implying that the service level of express outlets is not optimal, and needs to be further improved.

(2) The service area per outlet is decreasing.

Figure 4.11 illustrates the changes in the area of express service outlets in China from 2012 to 2017. It is evident that the service area per outlet is decreasing annually, yet the number of service outlets per 1,000 km^2 is going up. In 2012, the average number of express service outlets per 1,000 km^2 in China was 9.9, with one outlet providing coverage for an area of 100 km^2. Due to the extensive service coverage needed to meet the requirements of regional consumers, the service capacity failed to meet their satisfaction and the enterprise service quality was affected. By 2017, China had 22.5 express service outlets per 1,000 km on average, and the service area of each outlet was only 42 km^2. However, the density of outlets has increased and the service capacity has been enhanced.

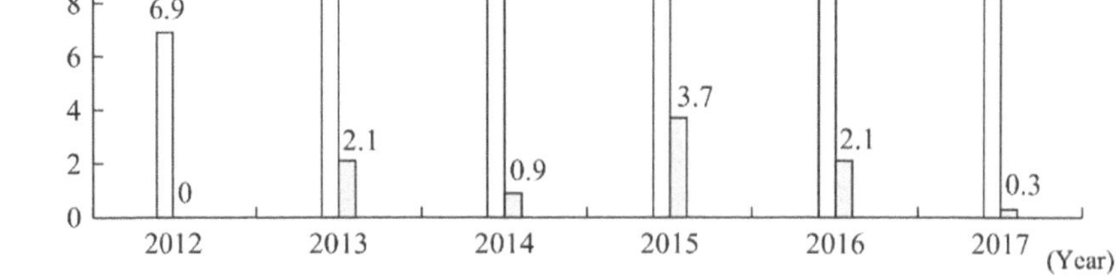

Fig. 4.10 Population density of national express service outlets and changes in the service. *Source* The Author

Fig. 4.11 Changes in the area density of China's express service outlets from 2012 to 2017. *Source* The Author

4.2.2.2 Uneven Distribution of Resources Among Nodes

The resources of nodes in both the fully-connected express delivery network and hub and spoke express delivery network structures are not distributed evenly. The east region has more resources, while central and western China have their own strengths and weaknesses. Table 4.4 shows the basic situation of express outlet services in 31 provinces across China in 2018. The eastern region, which includes Zhejiang province, Fujian province, Beijing and Guangdong province, has the highest population density. These four provinces all have an average of more than 20 express service outlets per 100,000 people, which is much higher than the average. Zhejiang has the highest number of express service outlets of all provinces, with 46 per 100,000 people. Conversely, Hebei has the lowest number, with only 13 per 100,000 people. The data of the western provinces are relatively stable, usually fluctuating around the average. And the data in central China fluctuates significantly. In conclusion, if we judge the service level of regional outlets based on the population they serve, eastern China has the highest service level, followed by western China, then central China.

Based on the density of express service outlets, the higher the density, the larger the business coverage. In this regard, eastern China has a significant advantage, as eight of the top ten regions are located in the east, namely Shanghai, Beijing, Zhejiang, Tianjin, Jiangsu, Guangdong, Shandong and Fujian. Of these, Shanghai has 1,214 outlets per unit area, far more than Beijing, which has 331. Central China has relatively stable data, while western China is significantly lower than average. This means that, in terms of the number of express service outlets, eastern China is ahead of the central, while the central is ahead of the west.

Table 4.4 The service level of express outlets in various regions in 2018 across the country

Province	Population density of express outlets (per/100,000 people)	Area of express service outlets (per/1,000 km^2)
Beijing	25	331
Tianjin	16	204
Hebei	13	52
Shanxi	18	43
Neimenggu	20	4
Liaoning	16	48
Jilin	17	25
Heilongjiang	16	13
Shanghai	20	744
Jiangsu	19	142
Zhejiang	46	245
Anhui	17	75
Fujian	26	83
Jiangxi	16	44
Shandong	15	93
Henan	15	85
Hubei	24	76
Hunan	15	49
Guangdong	22	140
Guangxi	16	34
Hainan	19	50
Chongqing	22	240
Sichuan	24	42
Guizhou	23	47
Yunnan	19	23
Xizang	35	1
Shaanxi	22	41
Gansu	18	11
Qinghai	18	2
Ningxia	22	23
Xinjiang	15	2
Average value	**20**	**97**

Source The Author

4.2.2.3 Distinct Control Capability of Nodes

In a fully-connected express delivery network structure, nodes have low control capability and little dependence on one another. Each operating node has a direct transport relationship with its subordinate nodes, resulting in a large number of transport routes that are difficult to manage. The carrying capacity between nodes is moderate. If a node or side of the network fails, it can easily be isolated.

In a hub and spoke express delivery network structure, nodes have a strong control capability and there is a high degree of interdependence between them. Each node only needs to connect to a particular hub central station in order to deploy, sort, and allocate packages. This enables effective control and management of the express's transportation direction.

By utilizing the Gephi statistical module to analyze the data set, the control capability of the nodes can be effectively identified. The results are depicted in Fig. 4.12. Nodes with a high control capability are considered to be critical nodes, as they can have a greater influence on the structure and function of the network than other nodes. The number of virtual nodes is minimal, but it can quickly spread and affect most network nodes. For example, when a key node of an express delivery network is deliberately damaged, it can cause the whole express delivery network to malfunction. Weibo is a vast network system, and users with many followers are key nodes whose tweets can quickly spread across the network, having a major influence. This demonstrates that critical nodes have a far-reaching impact on the network's structure and function.

Four metrics commonly used to measure the significance of nodes in complex networks are PageRank, Betweenness Centrality, Closeness Centrality, and Harmonic Closeness Centrality. These metrics are utilized to study the importance of multiple nodes.

The importance of each node is shown by its PageRank value, as seen in Fig. 4.12a. The higher the PageRank value, the more important the node is; the figure is sorted in descending order. From the figure, it is clear that "Chengdu Transit" is the most significant node in the data set, and is a major part of express transportation in the whole set.

The betweenness centrality of a node measures how many of the shortest paths between all pairs of nodes in a network pass through it. The higher the betweenness Centrality of a node, the more control it has over the other nodes in the network and the more important it is. Figure 4.12b shows the nodes in a network ranked in descending order of their betweenness centrality, with Guangzhou Center Station having the highest index.

Closeness centrality measures the average shortest distance from any node in a network to all other nodes, which reflects how quickly a node can be connected to other nodes. The shortest distance between nodes is the minimum number of nodes that must pass between any two nodes. Figure 4.12c displays the descending ranking of closeness centrality; the higher the index value, the greater the node's influence and range of service.

ID	Pagerank
Chengdu Transit Center	0.010081
Beijing	0.00973
Dongguan Center	0.008341
Guangzhou Center	0.008092
Shenzhen Center	0.008007
Bengbu Transit Department	0.007919
Xi'an Transit Center	0.007372
Wuhan Transit Department	0.006907
Shanghai	0.006875

(a)

ID	Betweenness Centrality
Guangzhou Center	10848.845238
Shenzhen Center	5959.728571
Dongguan Center	5249.975
Wuxi Transit Department	4395. 269048
Hangzhou Transit Department	4218.007143
Quanzhou Transit Department	4127.007143
Fuzhou Transit Department	3845.75
Wuhan Transit Department	3837.275
Zhengzhou Transit Center	3419.42381

(b)

ID	Closeness Centrality
Chengdu Transit Center	1.0
Bengbu Transit Department	1.0
Wuhu Transit Department	1.0
Shanghai Pudong Center	0.833333
Nanchang Transit Department	0.8
Hengyang Transit Center	0.8
Tianjin Transit Department	0.777778
Shenyang Transit Center	0.75
Xiamen Transit Department	0.666667

(c)

Fig. 4.12 The importance of node of express data set. **a** The importance of node under PageRank. **b** The importance of node of betweenness centrality. **c** The importance of node of closeness centrality. **d** The importance of node of Harmonic Closeness Centrality. *Source* The Author

ID	Harmonic Closeness Centrality
Chengdu Transit Center	1.0
Bengbu Transit Department	1.0
Wuhu Transit Department	0.91667
Shanghai Pudong Center	0.833333
Nanchang Transit Department	0.9
Hengyang Transit Center	0.880952
Tianjin Transit Department	0.75
Shenyang Transit Center	0.722222
Xiamen Transit Department	0.625

(d)

Fig. 4.12 (continued)

Harmonic closeness centrality is a measure of the distance from a given node to the most distant node in the network. The central node of the network is usually the one whose network radius is equal to its harmonic closeness centrality. As the harmonic closeness centrality of a node gets closer to the network radius, it becomes closer to the center of the network. Figure 4.12d shows a descending ranking of harmonic closeness centrality; it reveals that only 63.198% of the nodes in the express delivery network have a harmonic closeness centrality greater than 0, suggesting that most nodes are located near the edge of the network.

4.3 Characteristics of the Express Delivery Traffic Flow

Express delivery traffic flow is a measure of the current status of express transportation. It involves gathering, categorizing and assessing express logistics data over a specific time frame and determining how many express entities are passing through a certain point, section or lane of a road. Each express item is considered to be a unit of traffic flow, also known as an express traffic, and the aggregate of express traffic by express delivery companies over a period is known as express volume.

The express delivery traffic flow is characterized by vehicle operation, temporal and spatial dynamics, which can be random and vary depending on the time and location. The operation of express vehicles reflects the dynamic characteristics of the express transport process. The spatiotemporal distribution characteristics of the flow, which is the phenomenon of express delivery traffic flow changing with time and space, show the operation rules and characteristics of express traffic. The analysis of traffic flow can not only optimize the operations, networks, and transportation routes of express enterprises to meet customers' needs, but also improve their advantages.

4.3.1 Vehicle Operation Characteristics

4.3.1.1 Differentiation of Vehicle Operation Modes

Currently, there are two main operation modes for express delivery companies: direct operation and franchise systems (Wang et al., 2014). SF Express is the representative of direct operation, while companies such as STO Express, YTO Express, ZTO Express, BEST Express and Yunda Express use a combination of direct-operation and franchise modes. The operation and dispatching modes between transit centers and service outlets also differ accordingly.

Figure 4.13 shows the vehicle operation mode of direct-operated express enterprises. The headquarters manage and deploy express transport vehicles in a unified manner, and all express packages must be sorted and distributed by the transit center before being transferred to their next destination. The building, renovation, and expansion of the transit center are essential for domestic express enterprises to further expand their business scale and improve their network operation capabilities. The degree of direct operation of transit centers demonstrates the headquarters' management of the routing network, as well as their control over delivery efficiency and service quality. All companies are actively encouraging direct operation of these centers. Currently, there are 82 YTO Transit Centers (60 of which are directly operated), 82 STO Transit Centers (48 of which are directly operated), 74 ZTO Transit Centers (68 of which are directly operated), and 54 Yunda Transit Centers, all of which are direct-operation systems.

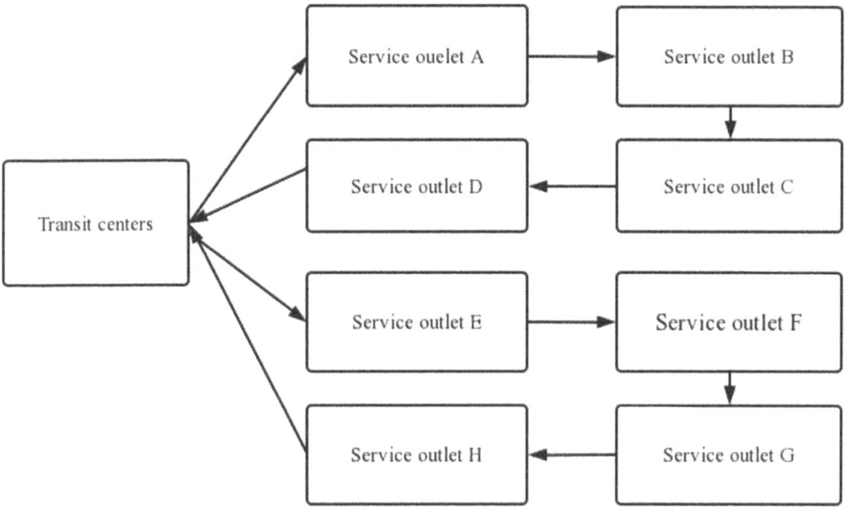

Fig. 4.13 Schematic diagram of vehicle operation of direct-operated express delivery enterprises. *Source* The Author

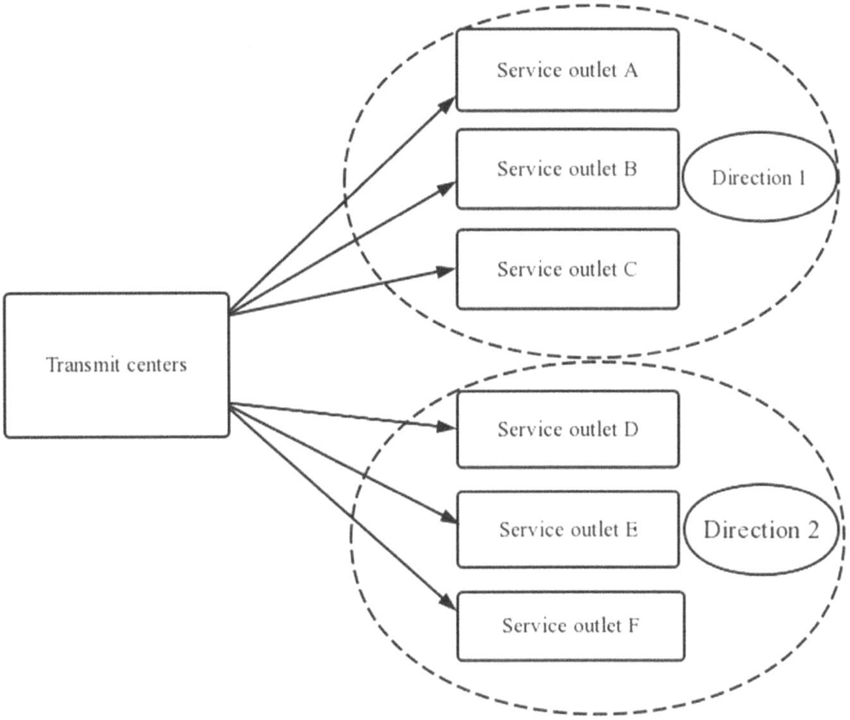

Fig. 4.14 Schematic diagram of vehicle operation of franchise express delivery enterprises. *Source* The Author

Figure 4.14 illustrates the operational mode of a franchise express enterprise, wherein the primary feature is distributing packages from service outlets to transit centers. The express transport vehicles are jointly managed by the headquarters and service outlets, with those from the latter traveling to the transit centers on a daily basis to pick up and deliver the express at a certain time.

To ensure timely delivery, some direct-operated express delivery companies, such as SF Express, have direct vehicle delivery systems. This system is similar to the franchise system in the last-mile delivery, with express vehicles used for delivery and box-type buses used for collection and delivery when the number or volume of the express is large.

4.3.1.2 Diversification of Vehicle Operation Time

(1) Vehicles entering and leaving the elevated road

As depicted in Fig. 4.15, there are four distinct peaks in the trips of express vehicles on the elevated road over the course of one day. From 6:00 a.m. to 7:30 a.m., vehicles of service outlets collect express packages from the distribution center. From 9:00

a.m. to 11:30 a.m., the city experiences its first delivery peak. From 1:00 p.m. to 2:00 p.m., there is the second delivery peak. From 6:00 p.m. to 10:00 p.m., vehicles of all service outlets take express packages to the distribution center and then come back.

We can use Xi'an as an example. Table 4.5 shows the traffic flow distribution. The road network of Xi'an is formed by two axes (five east–west roads and north–south streets), three ring roads (the first, second and third ring roads), and eight rays (Taiyi Road, Taibai Road, Xianning Road, Huaqing Road, Taihua Road, Zhuhong Road, Daqing Road, and Kunming Road). Currently, the length of the Southern Ring Road (which comprises of 10.79 km South Second Ring Road and 20.1 km South Third Ring Road) is shorter than that of the Northern Ring Road (10.47 km North Second Ring Road and 21.7 km North Third Ring Road). The Southern area (which includes the High-Tech Zone, South Residential Area, and Chang'an District, etc.) is the most economically developed area in Xi'an, with the highest population density and industrial clustering. Thus, it has become the most congested district of Xi'an. The northern part of the Ring Expressway has the most significant traffic flow of express vehicles, mainly because many private express delivery companies' distribution centers are located there and the expressways around the city are the main distribution channels to the city. Additionally, the South Second Ring Road and the East Third Ring Road also have relatively high traffic flow, with the South Second Ring Road hosting many express points and the East Third Ring Road being the primary channel for express vehicles to travel between the distribution center and the north and south of the city.

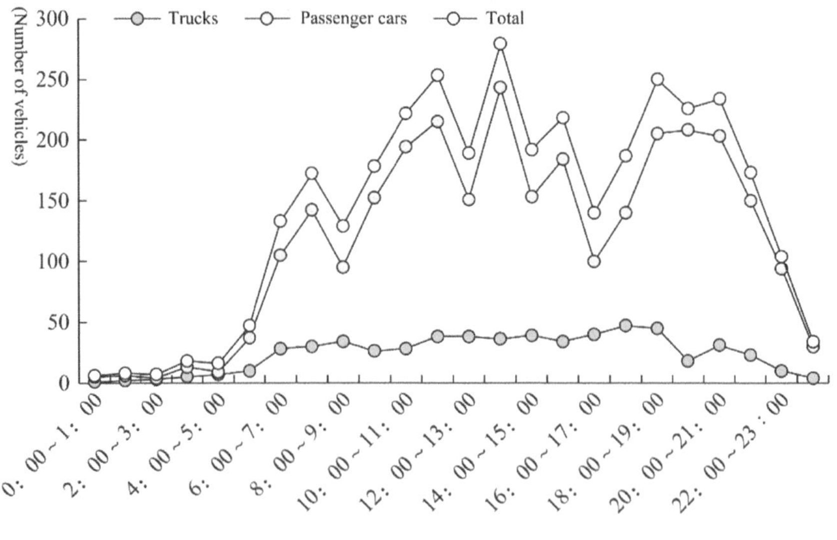

Fig. 4.15 The number of express delivery vehicles traveling on elevated roads by time period. *Source* The Author

Table 4.5 Traffic overview of Xi'an loop line

Loop Line	Length (km)	Year of opening
The Second Ring	30	2003
The Third Ring	74.8	2008
The Ring Expressway	88	2003
Outer Ring Expressway	270	2019

Source The Author

(2) Vehicles entering and leaving the central city

More than 80% of the service outlets in Xi'an are located within the Second Ring Road of the city, with the distribution center typically located outside the Outer Ring. Express enterprise vehicles travel to and from the distribution center at fixed times to pick up and deliver express packages, resulting in traffic flows both into and out of the central city. The distribution center of franchise express delivery companies has three pickup and delivery periods: 9:00–10:00 a.m. (first wave), 10:00–15:00 p.m. (second wave), and 18:00 to midnight (third wave). These periods are staggered to avoid the city's rush hour traffic. The headquarters manage the vehicles of direct-operated express enterprises in a unified manner to ensure timely delivery. Therefore, there is no fixed period for entering and leaving the central city, resulting in fewer vehicles entering and leaving the city with the franchise system compared to the direct management system. Moreover, the time for vehicles to enter and leave the city with the franchise system is less flexible than with the direct management system.

4.3.2 *Time Series Dynamic Characteristics*

Time-series dynamics can be divided into two benchmarks: long-term and short-term. Long-term characteristics refer to the expression of traffic flow in years and months. There are two main aspects to this.

4.3.2.1 Exponential Increase in Express Delivery Traffic Flow

We first analyze the distribution law of express traffic in terms of time series. We conduct the analysis from both the yearly and monthly time granularity. As shown in Fig. 4.16, the main ordinate in the figure represents the cumulative volume of express business, while the secondary axis shows the year-on-year growth rate of business volume. From 2010 to 2018, express business volume had increased for nine consecutive years. Traffic flow also rose annually in accordance with the increasingly mature macro development environment of the express delivery industry. Prior to

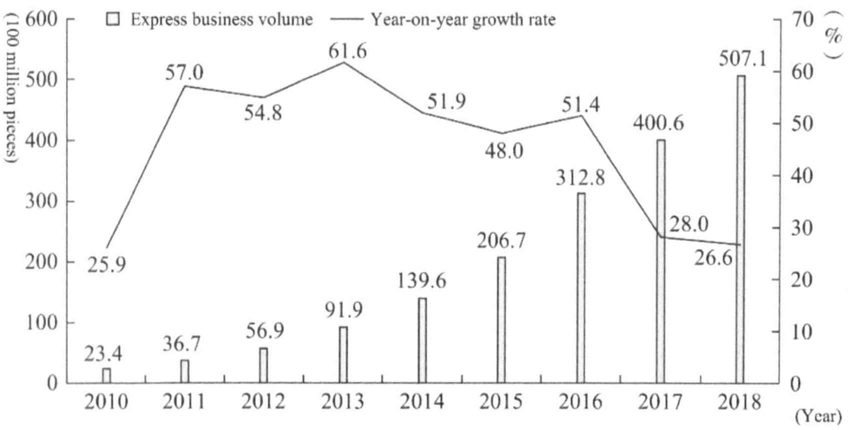

Fig. 4.16 The change trend of express delivery business volume from 2010 to 2018. *Source* The Author

2010, express business volume was lower than 2.5 billion pieces, with a low rate of growth. However, after 2011, it experienced a remarkable increase of. It surpassed 5 billion pieces for the first time in 2012, 20 billion pieces in 2015, and 50 billion pieces in 2018, amounting to 50.71 billion pieces in total, with a cumulative growth rate of 26.6%. Based on the current development environment, the volume of express business is expected to increase in the coming years, with a steady growth rate until it reaches the maximum capacity of e-commerce. Consequently, the traffic flow of the express delivery network will also increase.

The annual per capita express business volume change reflects the dynamic time characteristics of express delivery traffic flow at the micro level. As Fig. 4.17 shows, the secondary coordinates in the figure are the per capita express usage. It can be seen that from 2012 to 2018, there has been a continuous seven-year increase in per capita express usage, with a growing rate each year. The per capita express expenditure has an exponential upward trend. The micro-and macro-level conclusions both indicate the increasing trend of express delivery traffic flow.

4.3.2.2 The Annual Distribution Has a Three-Stage Structure

Figure 4.18 shows the distribution of traffic flow at a monthly time granularity. Comparing and analyzing the changes in express business volume from 2015 to 2018 reveals that express delivery traffic flow has a three-stage distribution with a peak in the middle, and declines at both ends, in a one-year time scale. In February, express business volume is the lowest. From March to November, express business volume steadily increased, with a slow rise from March to August and a significant jump from September to November. November saw the highest volume, then it decreased in December.

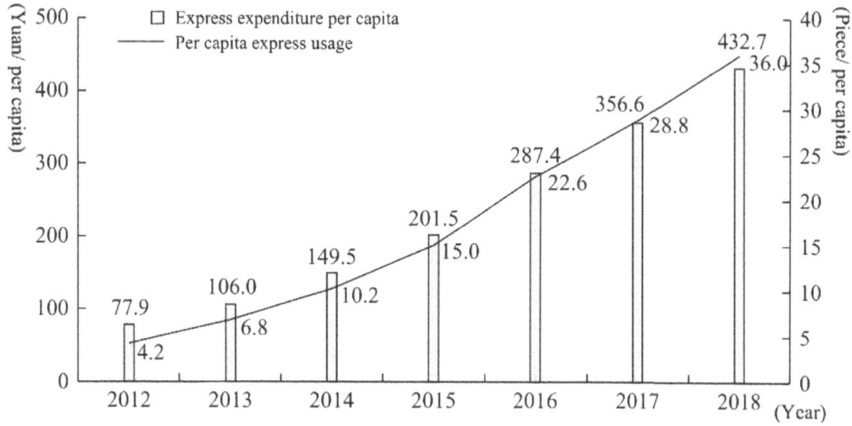

Fig. 4.17 The change trend of per capita express delivery business volume from 2012 to 2017. *Source* The Author

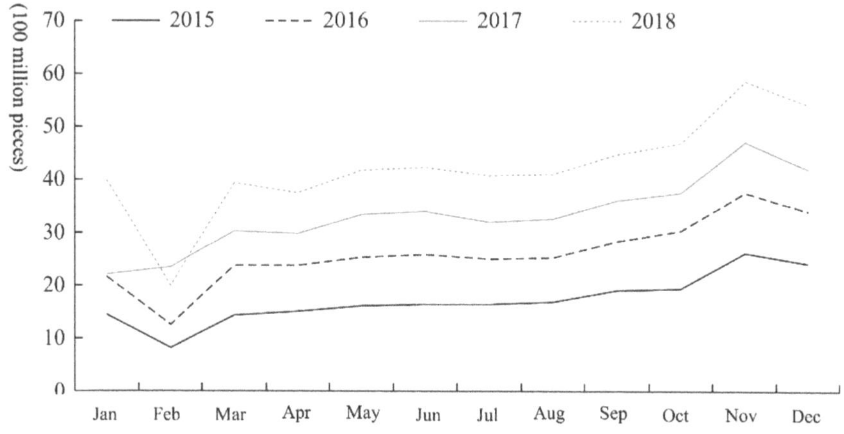

Fig. 4.18 The change trend of monthly express delivery business volume from 2015 to 2018. *Source* The Author

Express traffic is at its lowest in February due to the Spring Festival holiday, the longest holiday of the year. During this period, many express delivery companies stop delivering packages and some people are unable to sign for them due to trips. There are two reasons for the peak of express traffic in November each year: the subjective reason is the annual 'Double 11' express volume explosion as people tend to go on a shopping spree online; the objective reason is that express business volume is highly dependent on e-commerce in November.

Short-term traffic flow is divided into three stages: the free flow period (loose period), the transition period (general period), and the congestion period (peak period) of the day. Driving conditions vary between these periods:

(1) Free Flow Period

In this period, there are fewer express vehicles, so traffic flow is increasing slowly from the start, but remains relatively small. This allows vehicles to travel quickly without having to worry about other vehicles, reducing transportation time and ensuring goods are delivered on time.

(2) Transition Period

During this period, there are significantly more express vehicles than in the free flow period, and the density of express traffic is increasing. However, the speed of express vehicles is limited, so other vehicles on the road need to be taken into account. The traffic flow density will not continue to increase until it reaches a certain value, at which point express vehicles will reduce their speed and the traffic flow on the road will slow down. Therefore, when designing express transportation routes during this period, it is important to consider the overall traffic flow conditions on the road in order to select the routes with lower traffic flow density.

(3) Congestion Period

At this time, the traffic flow density has reached its peak. Vehicles will gradually slow down and the traffic flow will continuously decrease. With the increasing traffic flow density, vehicles are almost at a standstill during this period of congestion, and the traffic flow is nearly nonexistent. Heavy traffic causes delivery vehicles to be delayed for an extended period of time, which is not beneficial for the timely completion of delivery. Thus, delivery companies should take this into account when creating the initial route in order to avoid travelling to congested road sections at an inconvenient time.

4.3.3 Spatial Dynamic Characteristics

The spatial dynamic characteristics of express delivery traffic flow are explained from three levels: urban and rural distribution of express business volume, distribution on road sections, and regional distribution. To fully explore the spatial distribution characteristics of express delivery traffic flow, regional express business volume distribution will be analyzed in detail.

4.3.3.1 Urban and Rural Distribution

The imbalance between urban and rural economic development, production activities, and living standards has caused a significant difference in express delivery traffic flow between them. Rural roads have a much smaller flow of express business than

Table 4.6 Current status Quo of Roads in the Urban area of Xi'an, inside the Second Ring Road and the Xi'an circumvallation

Location	Main road		Secondary road		Road network	
	Length/km	Density (km/km^2)	Length/km	Density (km/km^2)	Length/km	Density (km/km^2)
The Xi'an Circumvallation	17.14	1.31	8.10	0.62	104.12	7.95
Inside the Second Ring	68.19	0.90	43.52	0.57	370.93	4.87
Road in urban areas	99.60	0.53	119.18	0.64	1009.6	5.40

Source The Author

urban roads due to the limited volume of express business in rural areas, but the rural express market is still promising. Currently, only EMS can deliver packages to villages, while other express delivery companies are setting up service outlets in towns, and recipients need to sign for the packages in person. SF Express and JD will contact the recipients beforehand. If a recipient doesn't want to sign for large items, like home appliances, by himself, the couriers will deliver the packages to his door. However, for some especially remote villages, delivery of one or two small packages will only be possible when the volume of express shipments in the same area reaches a certain amount.

4.3.3.2 Distribution on the Road Section

Due to the different levels, functions, and locations of each road section in the road network, the express delivery traffic flow of each section varies simultaneously. There are also significant differences in the express flow of urban entrances and exits. During the morning peak, the express traffic in the outbound direction is typically between 60 and 70%, while the proportion of express traffic entering the city is relatively high during the evening peak.

In Xi'an, the urban road construction follows the "two axes, three rings and eight lines" model. As shown in Table 4.6, the length of the Xi'an Circumvallation is much shorter than that of the urban area, but its density is the highest. This indicates that the roads in Xi'an have various differences, which should be taken into account when planning express transportation routes.

4.3.3.3 Regional Distribution

(1) The proportion of non-local express business volume continues to increase with the fastest growth rate.

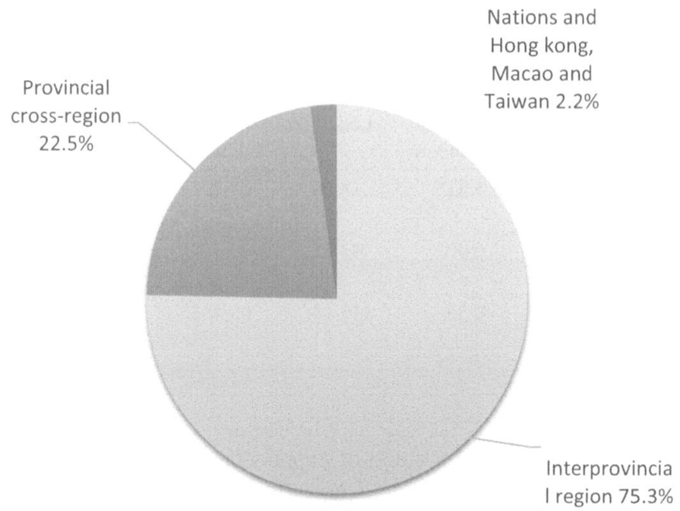

Fig. 4.19 Structure of express delivery business volume in 2018. *Source* The Author

Firstly, the structure and composition of express business volume are analyzed, as shown in Fig. 4.19. In 2018, interprovincial express accounted for 75.3% of the national express business volume, the highest proportion. Provincial cross-region express accounted for 22.5%, while international/Hong Kong, Macao and Taiwan express only 2.2%. Their agency revenue accounted for 67.5%, 19.7% and 12.7% of the total express revenue respectively.

In 2018, the provincial cross-region express volume totaled 11.41 billion pieces, a 23.1% increase from the previous year. The express agency revenue was 90.576 billion yuan, a 23.46% increase from the previous year. Compared with 2017, when the business volume and revenue experienced a 25% and 30% growth rate respectively, the growth rate significantly decreased.

Interprovincial express business remains dominant. Throughout the year, volume of interprovincial express packages was 38.19 billion, a year-on-year increase of 27.5%; the express business income was 310.37 billion yuan, a rise of 23.5%. The proportion continues to increase gradually, with 74.8% in 2017 and 75.3% in 2018.

As the social economy continues to develop, the amount of express business between nations and Hong Kong, Macao and Taiwan is also rising. In 2018, 1.11 billion pieces of international, Hong Kong, Macao and Taiwan express business were completed with an impressive growth rate of 34%, and the express business income reached 58.57 billion yuan, an increase of 10.4%.

(2) The proportion of express business differs from east to west in China.

As is shown in Figs. 4.20 and 4.21, Fast Unfolding layout is adopted in China, with the four top regional transit centers in terms of domestic express business volume being Guangzhou Transit Center, Chengdu Transit Center, Beijing Transit Center and Hangzhou Transit Center. The eastern region makes up the majority, while the western region makes up a relatively small proportion. The express delivery network in China covers most regions, and the routes are complex. There is a great disparity in the density of service outlets in different regions; the eastern region has a denser concentration, while the western region is relatively sparse. Furthermore, the number of hubs in the east is much greater than that in the west, making it possible to compare the level of social development and economic vitality between regions.

(3) The western express in China experienced the highest year-on-year growth rate in terms of both business volume and income.

In 2018, express business volume and income saw an increase in the eastern, central and western regions. In the eastern region, the express business volume for the whole year was 40.52 billion pieces, representing a year-on-year increase of 24.7%, while the express business income was 483.07 billion yuan, with a 20.5% year-on-year growth. The express business volume in the central region was 6.24 billion pieces, showing a year-on-year increase of 34.1%; the express business income amounted to 67.63 billion yuan, with a growth rate of 26.3%. For the western region, the express business volume totaled 3.96 billion pieces, an increase of 35.6% year-on-year; the express business income was 53.14 billion yuan, with a growth rate of 29.2%.

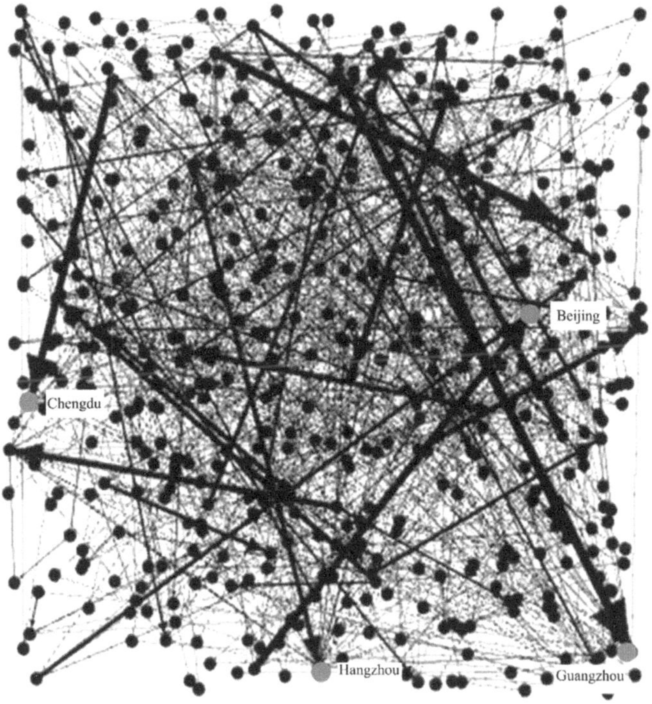

Fig. 4.20 Express delivery network topology under the layout of fast unfolding. *Source* The Author

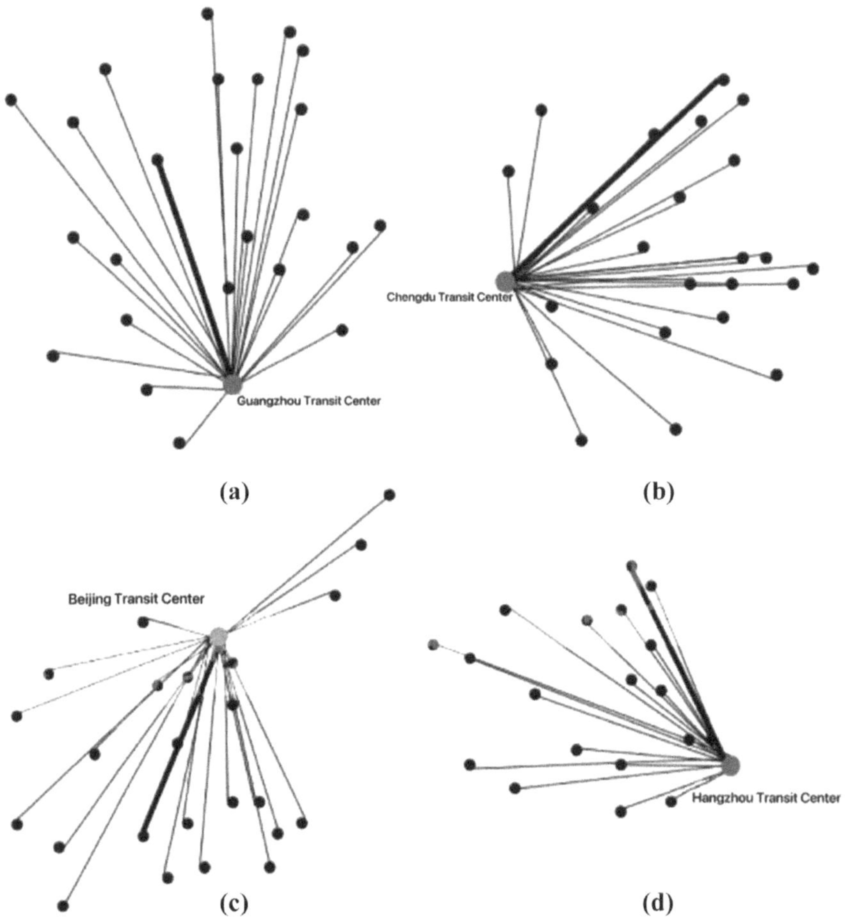

Fig. 4.21 Four regional transit centers. **a** Guangzhou Transit Center. **b** Chengdu Transit Center. **c** Beijing Transit Center. **d** Hangzhou Transit Center. *Source* The Author

In comparison to the eastern and central regions, the express business volume and income in the western region grew rapidly, becoming the fastest. This demonstrates that with the growth of the economic environment, residents of the western region are increasingly participating in the express delivery industry through e-commerce and other channels. Developing the western market will be essential for the future of the e-commerce and express delivery industry.

(4) Nearly 70% of the business volume is accounted for by the five provinces of Beijing, Shanghai, Guangdong, Zhejiang, and Jiangsu.

As shown in Tables 4.7 and 4.8, the total national express business volume in 2018 was 50.71 billion pieces, representing a 26.6% increase. The top five provinces and cities in terms of express business volume were Guangzhou, Zhejiang, Jiangsu, Shanghai, and Beijing, collectively accounting for 65.4% of the total.

The western region has the smallest business volume, yet the express business volume in certain provinces has seen a rapid rise. In 2018, seven of the eight provinces with the lowest express business volume were in the central and western regions. Among them, Ningxia had the highest year-on-year growth rate of over 80%, followed by Yunnan and Guangxi with increases of more than 40%, Guizhou, Neimenggu, Qinghai and Sichuan with increases of more than 30%, and Gansu, Xinjiang and Shaanxi with increases of more than 20%. Xizang, in particular, recorded a 27.9% growth rate—a significant improvement from the −22.7% it experienced the year before. Although the western region lags behind in the scale of industry development and economic level, the express volume has risen substantially.

The central provinces have experienced an overall good development momentum, with the year-on-year growth rate of express business surpassing 20%. However, the business volume remains low.

Table 4.7 Express business volume and per capita ranking in some regions of My China in 2007 and 2018

2007 total		2018 total		2007 per capita		2018 per capita	
Region	Rank	Region	Rank	Region	Rank	Region	Rank
Shanghai	1	Guangdong	1	Shanghai	1	Jinhua	1
Guangdong	2	Zhejiang	2	Guangdong	2	Guangzhou	2
Beijing	3	Jiangsu	3	Beijing	3	Hangzhou	3
Jiangsu	4	Shanghai	4	Jiangsu	4	Shenzhen	4
Zhejiang	5	Beijing	5	Zhejiang	5	Dongguan	5
Shandong	6	Fujian	6	Shandong	6	Shanghai	6
Fujian	7	Hebei	7	Fujian	7	Beijing	7
Sichuan	8	Henan	8	Sichuan	8	Suzhou	8
Hebei	9	Sichuan	9	Hebei	9	Jieyang	9
Henan	10	Hubei	10	Henan	10	Jiaxing	10
Liaoning	11	Anhui	11	Liaoning	11	Taizhou	11
Hubei	12	Hunan	12	Hubei	12	Qaunzhou	12
Tianjin	13	Liaoning	13	Tianjin	13	Ningbo	13
Hunan	14	Jiangxi	14	Hunan	14	Wenzhou	14
Heilongjiang	15	Tianjin	15	Heilongjiang	15	Nanjing	15
Shaanxi	16	Shaanxi	16	Shaanxi	16	Wuxi	16
Anhui	17	Guangxi	17	Anhui	17	Wuhan	17
Jiangxi	18	Yunnan	18	Jiangxi	18	Chengdu	18
Yunnan	19	Shanxi	19	Yunnan	19	Zhengzhou	19
Chongqing	20	Heilongjiang	20	Chongqing	20	Tianjin	20

Source The Author

Table 4.8 Business volume and income of express delivery companies in some provinces in 2018

Province	Express delivery agency volume (10,000 pieces)	Year-on-year growth (%)	Express delivery income (ten thousand yuan)	Year-on-year growth (%)
Country	5,071,042.8	26.6	60,384,253.8	21.8
Beijing	220,875.6	−2.9	3,310,328.2	9
Tianjin	57,576.7	14.7	874,555.6	14.6
Hebei	174,136.2	45.9	1,807,779.7	42.9
Shanxi	30,332.9	24.5	385,467.2	28.5
Neimenggu	15,182.3	37.6	299,100.7	24.8
Liaoning	65,363.7	27.1	879,726.8	29.3
Jilin	22,637.5	28.8	377,131.5	23.9
Heilongjiang	30,177.2	30.2	464,325.9	29.5
Shanghai	348,648.8	11.9	10,202,806	17.4
Jiangsu	438,935.4	22.1	4,808,932.1	17.8
Zhejiang	1,011,050.7	27.5	7,793,024.1	16.6
Anhui	112,322.4	30.1	1,110,120	23.9
Fujian	211,613.4	27.4	2,066,842.1	27.6
Jiangxi	61,929.5	41.5	670,877.1	36.4
Shandong	218,701.1	44.4	2,283,960.8	33.9
Henan	152,631.6	42.1	1,529,449.7	31.9
Hubei	135,307.7	33.6	1,437,737.2	20.8
Hunan	78,932.6	33.4	804,691.3	25.4
Guangdong	1,296,195.7	27.9	14,117,279.4	23.1
Guangxi	48,101.1	51.5	615,000.7	37.1
Hainan	7107.1	20.1	163,135.3	28.5
Chongqing	45,795.4	39.3	580,358.4	29.7
Sichuan	145,991.7	31.8	1,671,575.4	31.1
Guizhou	21,193.7	34.3	404,531.8	29.9
Yunnan	33,999.1	49.3	471,438.1	30.9
Xizang	725.8	27.9	24,297.1	18.6
Shaanxi	56,876.5	24.3	673,078.9	19.4
Gansu	8911.6	23.7	188,515.3	27.3
Qinghai	1897.2	30.9	47,871.9	23.3
Ningxia	6771.3	82	81,303	19.9
Xinjiang	11,121.4	23	239,012.4	26.1

Source The Author

From an income point of view, the Eastern region is the driving force behind the express delivery industry. In 2018, the top 8 provinces in terms of express business revenue were all located in the Eastern region and they accounted for more than 78% of the total express business volume.

(5) Cities in Guangdong, Zhejiang and Jiangsu and those classified as first-tier cities, have a high volume of express agencies.

Table 4.9 shows that the top five cities in terms of express business volume in 2018 were mainly first-tier cities, such as Guangzhou, Shanghai, Shenzhen and Hangzhou. Furthermore, 15 first-tier cities, including newly-established cities, were listed among the top 35. Apart from these first-tier cities, most of the top 50 cities are located in Guangdong, Jiangsu and Zhejiang provinces. For instance, cities like Dongguan, Guangzhou, Jieyang, Huizhou, Foshan, Shantou and Zhongshan are all in Guangdong Province; Nantong, Xuzhou, Changzhou, Suqian, and Yangzhou are in Jiangsu Province; and Jinhua, Wenzhou, Taizhou, Jiaxing, Shaoxing and Huzhou are in Zhejiang Province. This is consistent with the provincial express ranking.

(6) A few regions carry the majority of express flow.

The express business is heavily concentrated in a few cities, as evidenced by Fig. 4.22, which shows that the top eleven cities in terms of express business volume (Guangzhou, Jinhua, Shanghai, Shenzhen, Hangzhou, Beijing, Dongguan, Suzhou, Chengdu, Quanzhou and Jieyang) account for over 50% of the total express business volume. This indicates that the express delivery traffic flow is highly concentrated spatially, with a few regions accounting for the majority of the express business and the eastern region taking the lead.

To summarize, the express delivery traffic flow exhibits specific typical characteristics in terms of spatial and temporal dynamics: firstly, the express delivery traffic flow increases exponentially over time; secondly, the dynamic spatial trend of the express delivery traffic flow is concentrated, with a few areas dominating.

4.4 Summary

This chapter sorts out the structure and characteristics of the express delivery network. To do so, it draws on the findings of previous scholars to construct the express delivery network topology and sorts out its structure based on the components of the network. Analysis reveals two different express delivery network structures: a fully-connected express delivery network and a hub and spoke express delivery network, which demonstrate the characteristics of a small world without scale. Next, this chapter examines the structural features of the express delivery network and delves deeper into the regional layout of the express delivery network. It highlights the

Table 4.9 The top fifty cities in terms of express delivery business volume in 2018

Ranking	City	Express delivery business volume (10,000 pieces)	Ranking	City	Express delivery business volume (10,000 pieces)
1	Guangzhou	506,447.8	26	Changsha	44,408.0
2	Jinhua	366,123.2	27	Jinan	43,195.5
3	Shanghai	348,648.8	28	Shaoxing	41,993.7
4	Shenzhen	320,825.6	29	Fuzhou	40,904.9
5	Hangzhou	258,910.0	30	Xi'an	39,944.9
6	Beijing	220,875.6	31	Nantong	39,362.0
7	Dongguan	133,853.6	32	Qingdao	38,323.4
8	Suzhou	124,563.0	33	Zhongshan	36,025.8
9	Chengdu	104,785.4	34	Baoding	35,530.9
10	Quanzhou	97,247.4	35	Linyi	33,324.6
11	Jieyang	95,962.0	36	Shenyang	31,208.1
12	Wuhan	92,636.3	37	Xiamen	30,310.2
13	Wenzhou	91,485.0	38	Huzhou	30,069.4
14	Ningbo	78,474.8	39	Nanchang	27,718.2
15	Nanjing	76,634.5	40	Huizhou	27,408.7
16	Taizhou	70,312.4	41	Xuzhou	26,250.9
17	Zhengzhou	68,246.3	42	Nanning	25,097.6
18	Tianjin	57,576.7	43	Langfang	24,482.2
19	Shantou	57,480.8	44	Changzhou	22,811.4
20	Jiaxing	53,787.6	45	Kunming	22,367.4
21	Shijiazhuang	52,268.3	46	Harbin	21,684.8
22	Wuxi	51,181.9	47	Suqian	20,800.4
23	Foshan	47,657.0	48	Weifang	18,902.2
24	Hefei	47,287.7	49	Yangzhou	15,459.5
25	Chongqing	45,795.4	50	Dalian	15,349.3

Source The Author

uneven distribution of the express delivery network infrastructure. Finally, the characteristics of express delivery traffic flow are analyzed. Results show that express delivery traffic flow on an annual time scale has a three-stage distribution characteristic with a rapid increase year over year. Furthermore, the express delivery traffic flow varies significantly among regions, with a few regions carrying the majority of it. By combining the structural characteristics of express delivery networks with the level of regional economic development, the reasons for the differences in traffic flow are identified.

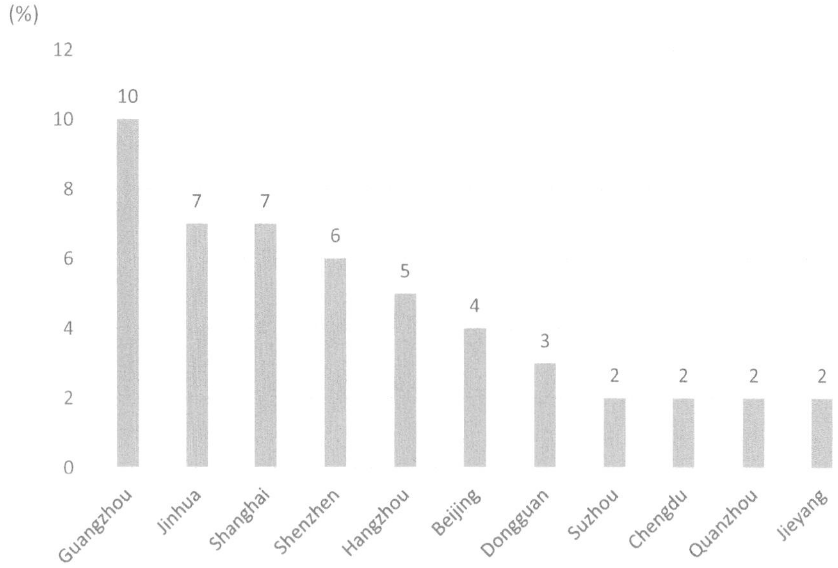

Fig. 4.22 Cumulative frequency of express delivery business volume. *Source* The Author

References

Bai, M. G., & Zhu, J. F. (2006). Comparative study on fully-connected and hub-and-spoke airline network. *System Engineering Theory and Practice, 26*(9), 113–117.

Chen, Y., Du, J., Sherman, H. D., & Zhu, J. (2010). DEA model with shared resources and efficiency decomposition. *European Journal of Operational Research, 207*(1), 339–349.

Li, Y. (2006). *Research on Hub and Spoke Network Theory and its application* [Doctoral Dissertation, Fudan University].

Lin, T., & Xie, X. C. (2017). Elements, Structure of express delivery network and their spatial pattern in Shanghai. *Transport Research, 3*(5), 9–49.

Ni, L. L. (2010). Analysis of the Characteristics of hub-and-spoke, point-to-point, and combined express networks. *Statistics and Decision, 20*, 59–61.

Wang, Z. Q., Zhu, H., & Qin, M. X. (2014). *New urbanization and transportation development—Proceedings of the Annual Meeting and 27th Symposium on Urban Transportation Planning in China in 2013* [Conference]. Urban Transportation Planning Academic Committee of China Urban Planning Society, Beijing, China.

Yang, H. R., Gu, L. Y., & Zheng, J. (1996). *Organization and management of postal communication network*. Beijing University of Posts and Telecommunications Press, 5–26.

Zhang, H. B., & Nie, Y. C. (2009). Analysis on network system of china express delivery enterprises. *Logistics Technology, 28*(9), 35–37.

Chapter 5
Express Delivery Network Optimization Based on Intelligent Algorithms

The optimization of express delivery networks refers to the overall optimization of the distribution routes between the various outlets of a physical layer network under the constraints of the management network. It has been found, based on the characteristics of "time-limited delivery" and "low-carbon operation" of express delivery enterprises, that the distribution paths and carbon emissions between nodes are related to the structure of express delivery network, thus implying that a change of its structure will affect the distribution cost and environmental cost. Additionally, intelligent algorithms are of great significance in the large-scale network optimization. Thus, the optimization of express delivery network through intelligent algorithms can effectively address the problem of express delivery routes.

5.1 Overview of Network Optimization Algorithms

5.1.1 Introduction of Basic Algorithms

Currently, people are able to utilize a variety of algorithms for network optimization, which generally fall into two categories: Accurate algorithms and heuristic algorithms (Huang, 2010).

5.1.1.1 Accurate Algorithm

Accurate algorithms refer to methods which can obtain the optimal solution, commonly including the network flow algorithm and dynamic programming technique. The characteristic of accurate algorithms is that the amount of computation is positively correlated with the size of the optimization object. Therefore, this algorithm is suitable for solving small-scale optimization problems. Furthermore, due to

P. Li and J. Wu, *Research on Innovation of Express Delivery Network Management*,
https://doi.org/10.1007/978-981-99-8982-9_5

the fact that the accurate algorithms are derived mathematically, they are prone to the exponential explosion problem. Thus, heuristic algorithms are often utilized to reduce computation time in practical applications (Song & Song, 2003).

5.1.1.2 Heuristic Algorithm

Heuristic algorithms utilize inductive reasoning based on observations and analysis of real-world experiments and experiences to identify suitable models and algorithms for a given problem. After applying heuristic information to the determined feasible solution, further iterations may lead to the optimal feasible solution. Most network optimization problems in the real world are classified as Non-deterministic Polynomial (NP) problems, which are large in scale and difficult to solve for an optimal solution in a short period of time, thus rendering heuristic algorithms to be an attractive option. Heuristic algorithms can be classified as follows:

1. Traditional Heuristic Algorithm

Commonly, traditional heuristic algorithms include a two-stage method, a mileage-saving method, a scanning method, and so on These algorithms, as they start from a single point, often require a significant amount of time to reach the optimal gradient information in the entire network. Additionally, due to their limited specificity and the complexity of constraints, these algorithms could prove more effective in solving problems such as practical applications of localization, economic optimization, structural optimization, and engineering design.

2. Modern Heuristic Algorithm

Modern heuristic algorithms, also referred to as meta-heuristic algorithms, are commonly utilized in engineering due to their four key characteristics: (1) simplicity, flexibility and ease of implementation; (2) randomness without prioritized or gradient information; (3) effective avoidance of local optima; (4) lack of subject restriction.

The meta-heuristic algorithm is based on the principle of building algorithm models based on biological or physical phenomena from nature in order to solve complex optimization problems in reality. This approach is typically divided into four parts, which are evolutionary algorithms, physical algorithms, population algorithms, and human algorithms, as depicted in Fig. 5.1. The laws of evolution motivate the development of evolutionary algorithms in nature. This process begins with a randomly constructed population that progresses through successive generations. An advantage of this technique is that the best individuals from one generation are combined to create the following generation; this allows the population to be improved iteratively. The most widely used evolutionary heuristics are Genetic Algorithms (GA), Simulation of Darwin's Theory of Evolution (Holland, 1975), Evolutionary Strategies (ES) (Rechenberg, 1978), Probability-based incremental learning (PBIL) (Dasgupta & Michalewicz, 1997), Genetic Programming (GP) (Koza, 1992), and Differential Evolution (DE), among others.

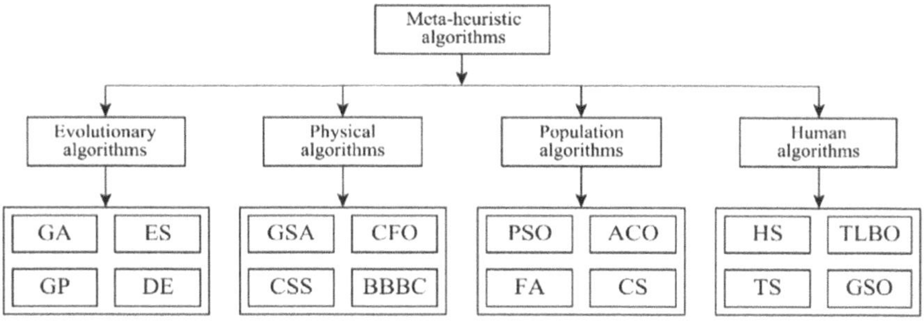

Fig. 5.1 Classification of meta-heuristic algorithms. *Source* The Author

Physical algorithms simulate the physical laws of the universe. The most well-known algorithms include Simulated Annealing (SA) (Amagai et al., 1983) (ErnÝ, 1985), Gravitational Search (GSA) (Webster & Bernhard, 2003), Central Gravitational Optimization (CFO), Charged System Search (CSS), and Big Bang Algorithm (BBBC).

Population algorithms are a type of computational technique that models the behavior of animals. The most renowned of these is Particle Swarm Optimization (PSO) (Kennedy et al., 1995), which is based on the social behavior of birds. The core concept of PSO is to utilize the motion of particles within the search space to identify the best solution. Another well-known example is the Ant Colony Optimization (ACO) (Dorigo et al., 2006) which mimics the behavior of ants to discover food and the shortest route, followed by the Firefly Algorithm (FA) and Cuckoo Search Algorithm (CS).

Human algorithm is a novel algorithm developed through the mathematical deduction and synthesis of existing algorithms. Common forms of this algorithm include the Harmony Search Algorithm (HS), Teaching–Learning-Based Optimization (TLBO), Tabu Search Algorithm (TS), and Group Search Algorithm (GSO).

The advantages and disadvantages of common heuristic algorithms are shown in Table 5.1.

5.1.2 *Selection of Algorithm*

5.1.2.1 The Criteria for Choosing Ant Colony Optimization

Ant Colony Optimization (ACO), also known as the ant algorithm, is used to find the optimal path. Initially proposed by Marco Dorigo in 1992, it is derived from the process of ants actively seeking the most efficient route when foraging for food. As an evolutionary imitation algorithm, the ant colony optimization is advantageous in many respects. By comparison with other algorithms, it has been found that the

Table 5.1 Advantages and disadvantages of common heuristic algorithms

Name	Advantages	Disadvantages
Genetic algorithm	Wide coverage is conducive to global optimization. Multiple individuals in the group can be processed in parallel, reducing the risk of falling into optimal local solutions. The random search is efficiently	There is a difficulty in parameter optimization. It is easy to converge prematurely, and the local search ability is weak
Tabu search algorithm	In the iterative solution process, inferior solutions can be accepted, and the local detection ability is strong	The global exploration ability is poor, and the calculation time is long. It cannot be guaranteed that the optimal solution is sought
Simulated annealing algorithm	The solution speed is fast. The initial value has strong robustness, simple, universal and easy to implement	The optimization process is long. It is suitable for solving small-scale problems
Ant colony optimization	It has a strong parallel search ability, and each individual has information exchange, which can promote the optimal solution	The algorithm is complex and requires a long search time. It is prone to stagnation during optimization. It is easy to fall into local optimum
Particle swarm optimization	The algorithm is simple, and the search speed is fast. It is suitable for dealing with real-valued problems	Easy to fall into local optimum; not suitable for discrete problems
Bat algorithm	It has few parameter settings, simple operation and short running cycle, which is convenient for solving large-scale VRP problems	It is easy to fall into local optimum and premature convergence
Firefly algorithm	Few parameters, easy to implement, high efficiency, strong global search ability	Usually, there will be multiple peaks and multiple poles, which will converge prematurely and fall into local optimum

Source The Author

ant colony optimization is more precise and effective than other algorithms when it comes to parameter optimization (Zhang & Shao, 2008).

Ant colony optimization has several characteristics, including self-organization, parallel search capabilities, positive feedback, and robustness.

1. Self-organization

In Systems Engineering, there are two main categories of organizations: self-organization and organization by external entities. The distinction between them lies in the source of their organization, whether it is internal or external to the system. Self-organizing instructions are generated from within the system, while the latter originate from outside the system. A system is said to be self-organizing when there is no external interference in its acquisition of temporal and spatial characteristics.

Self-organization implies that the process of entropy reduction in a system occurs autonomously, with no external intervention. The ant colony optimization is the best example of this. Initially, ants search for a solution in a disordered manner. However, after numerous iterations and cycles, the shortest path information is shared between ants through pheromone, thus encouraging more ants to move towards the optimal path. The process of finding the optimal path is thus disordered at the outset, but orderly at the end.

2. Parallel Search Capability

The Ant Colony Optimization is part of a multi-level system as the path-finding process is independent for each ant and relies upon pheromones to transfer information between them. When beginning the search for an optimal path, each ant performs an independent search at various points simultaneously which gives the algorithm certain reliability and powerful global search capabilities.

3. Positive Feedback

In the food-finding process of ants, it can be observed that the reason why ants can locate the shortest path is largely dependent on the concentration of pheromones on the optimal path. The increase in concentration is a positive feedback process. Initially, the pheromone concentration on each path is identical and any minor fluctuations within the system will affect the pheromone concentration on the path. The algorithm principle is such that subsequent ants are drawn towards the path with the highest concentration of pheromones, which is the optimal path. This positive feedback process can continually raise the pheromone concentration on the optimal path until the system eventually evolves towards the optimal solution. The positive feedback characteristic is the essential element of the ant colony optimization and it is this precise feature that allows the algorithm to continue to evolve.

4. Robustness

In comparison with other algorithms, ant colony optimization does not necessitate an initial route due to its ability to independently carry out optimal routes without relying on the initial route, and manual intervention is not required during the process of finding optimal routes. Additionally, its parameter setting is straightforward, which enables it to effectively address other combinatorial optimization problems.

5.1.2.2 Rationale for Selecting Simulated Annealing

Simulated Annealing (SA) is a heuristic algorithm inspired by the annealing process in thermodynamics, which is used to extend local to global searches in order to approximate an optimal solution. The algorithm iteratively reduces the temperature parameter and selects the neighboring state with the lowest target value with a given probability formula in order to find the solution, thus making it a theoretically comprehensive optimization technique.

SA is adept at solving optimization problems that many traditional algorithms are unable to, such as combinatorial optimization problems and NPC problems. This advantage can be leveraged to solve problems with multi-dimensional and complex characteristics. Moreover, SA can be used to optimize non-differentiable or discontinuous functions, rendering it highly efficient and reliable to obtain high-quality solutions.

The main advantages of SA are its ability to find a comprehensive optimal solution and its fast convergence. By conducting multiple iterations, the current optimal solution is obtained. The essence of SA is the outer circulation which effectively maintains temperature adjustment and guarantees that the solution when the temperature reaches the settings is the global optimum. When the initial temperature is high enough and its falling trend is slow, the algorithm performs a global search which utilizes the Metropolis sampling criterion to continuously select or discard a series of new solutions, increasing the likelihood of obtaining the global minimum solution. Generally, compared to other algorithms, SA is efficient in solving problems, has a wide application, is simple to operate, has implicit parallelism, and does not impose any constraints on the objective function.

5.2 Optimizing Express Delivery Networks with Ant Colony Optimization

5.2.1 Ant Colony Optimization (ACO)

5.2.1.1 The Mode of ACO

1. State Transition Probability Mode

$$p_{ij}^{k}(t) = \begin{cases} \frac{\tau_{ij}^{\alpha}(t)\eta_{ij}^{\beta}(t)}{\sum_{r \in allowed_k}\tau_{ij}^{\alpha}(t)\eta_{ij}^{\beta}(t)}, & j \in allowed_k \\ 0, & otherwise \end{cases} \quad (5.1)$$

α—heuristic factor, which indicates the importance of the previous path, reflects the probability that chooses other ants' path in the subsequent cycle.
β—expectation factor, which indicates the importance of visibility, determines the degree to which heuristic information is valued in the process of ants finding their path.
$allowed_k$—the cities that ant k can pick in the next cycle.

2. Pheromone Update Mode

$$\tau_{ij}(t+1) = (1-\rho)\,.\,\tau_{ij}(t) + \Delta\tau_{ij}$$
$$\Delta\tau_{ij} = \sum_{k=1}^{m} \Delta\tau_{ij}^{k} \tag{5.2}$$

Among them:

$$\Delta\tau_{ij}^{k} = \begin{cases} Q/L_k,\ Antkpassesthroughcitiesiandjinthiscycle \\ 0,\ otherwise \end{cases} \tag{5.3}$$

ρ—pheromone volatility coefficient, $(1-\rho)$– pheromone residual factor;
$\Delta\tau_{ij}^{k}$—the pheromone increments on path (i, j) during the cycle, with an initial value of 4;
$\Delta\tau_{ij}^{k}$—the amount of pheromone left by the ant K on path (i, j);
Q—pheromone concentration, which affects the convergence rate of the algorithm.
L_K—the length of the distance traveled by the ant K during the cycle.

5.2.1.2 Ant Colony Optimization Solution Approach

The most common application of Ant Colony Optimization in the optimization of a logistics network is the Travelling Salesperson Problem (TSP), in which the shortest path is determined through a series of iterations.

When employing ACO to address TSP, it is postulated that N ants can traverse between nodes only. The transition probability of each ant is calculated based on two parameters: the first is the information quality, otherwise referred to as the pheromone trail, and the second is the degree of visibility, also known as the prior value.

There are two primary reasons for pheromone renewal. The first is volatilization, which refers to the gradual decrease in pheromone concentration along the path, similar to the volatilization of substances in nature. The second is reinforcement, in which the concentration of pheromone is higher on paths that many ants have traveled.

The ants in the subsequent stage accomplished the action goal through "roulette", which essentially signifies that the probability of accessing the subsequent corresponding node is determined by the data on the current node. Following multiple iterations, the optimal solution can be incrementally approached.

Whenever an ant finds a solution, its degree of optimization is automatically evaluated and the evaluated information is stored in the corresponding solution space simultaneously. The specific algorithm flow is depicted in Fig. 5.2.

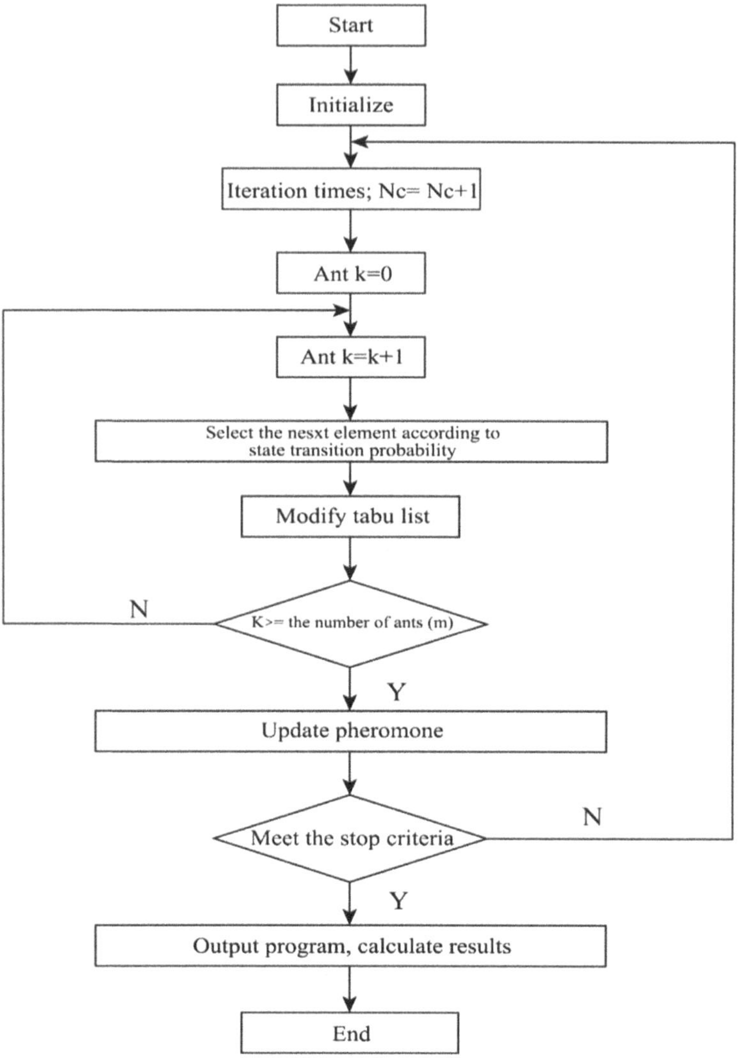

Fig. 5.2 Flow of ant colony optimization. *Source* The Author

5.2.1.3 Enhancement of Ant Colony Optimization

The Ant System with Elitist Strategy is the earliest application of Ant Colony Optimization (ACO) for improvement. This algorithm is characterized by the fact that even after finding the optimal solution, other ants still continue to look for it, thus increasing the amount of pheromone associated with that solution until the best possible solution is obtained. This improved algorithm is beneficial in accelerating convergence. However, there are some disadvantages. For instance, after initially finding the optimal solution, the ants tend to stay around it. This reduces the variety of solutions, which affects the subsequent selection probability, thus driving the evolutionary process away from the optimal solution and impeding the exploration of it.

The Max–Min Ant System has the characteristic of fully utilizing the cycle of finding the optimal solution. In each iteration, the system only updates the pheromone; thus, the optimal solution may be obtained either in the current iteration or after the iteration is completed. During the whole process of optimization, the pheromone updating process will be performed on the path of each ant in order to prevent the search from terminating.

The enhancement of the Best–Worst Ant System not only bolsters the optimization capacity of ACO in the global process, but also guides the concept of ACO to develop in the direction of optimization. This improved algorithm was designed to increase the weight of the best solution and reduce the weight of the worst solution. Furthermore, it can intensify the disparity in pheromone concentration between the optimal path and the worst path, which can help ants to more easily find a path closer to the optimal solution.

5.2.2 *Analysis of Simulation Results*

5.2.2.1 Problem Description

It is assumed that a postal distribution center delivers mail to multiple demand outlets and that all vehicles must return to the distribution center after completing the delivery task. The locations and mail volumes of all demand outlets are known, and the load capacity of each vehicle is specified. The configuration of the default postal vehicle is the same and there is a sufficient number of postal vehicles. It is required to arrange the travel routes of postal vehicles rationally so as to minimize the total distance (lowest freight costs), carbon footprint, and a number of vehicles. During the journey, regardless of external influence, the postal vehicle runs at a constant speed of 70 km/h. The distance between each checkpoint can be considered as the cost; that is, the cost problem is transformed into a postal route problem.

The three main automobile lines of Shaanxi Provincial Post, namely, Xi'an to Tongguan, Xi'an to Luochuan, and Xi'an to Hancheng are used as examples, as shown in Table 5.2.

The Shaanxi Postal Network Dispatching Center starts from Xi'an and delivers mail to Hancheng, Tongguan and Luochuan respectively. Table 5.3 shows the distances (in kilometers) and coordinates of the fourteen checkpoints along the route, based on the 1:900,000 Shaanxi Province map published in 2007. The origin of the coordinates is the lower left corner of the map and the unit is centimeters. The distribution has been plotted in Matlab, as shown in Fig. 5.3.

The carbon emission cost primarily references the emission cost, and during the mail distribution process, it primarily denotes the emission amount generated by the

Table 5.2 Basic situation of three postal routes in shaanxi province

Name	Order	Total Mileage (km)
Xi'an—Tongguan	Xi'an → Weinan → Huaxian → Huayin → Tongguan → Xi'an	367
Xi'an—Luochuan	Xi'an → Yaozhou → Tongchuan → Yijun → Huangling → Luochuan → Xi'an	471
Xi'an—Hancheng	Xi'an → Heyang → Ruyi → Longting → Zhichuan → Hancheng → Xi'an	534

Source The Author

Table 5.3 Coordinates and distances of checkpoints

Number	Name	X-axis	Y-axis	0	1	2	3	4	5	6	7
				Xi'an	He yang	Ruyi	Longting	Zhichuan	Hancheng	Weinan	Huaxian
0	Xi'an	413	337	0							
1	Heyang	534	456	234	0						
2	Ruyi	534	465	254	14	0					
3	Longting	545	470	264	20	10	0				
4	Zhichuan	560	474	271	40	17	7	0			
5	Hancheng	562	485	286	55	32	22	15	0		
6	Weinan	470	365	81	130	143	164	157	173	0	
7	Huaxian	495	367	140	130	144	164	159	173	27	0
8	Huayin	530	375	178	162	176	196	191	205	60	38
9	Tongguan	545	372	205	190	204	149	219	233	88	66
10	Yaoxian	415	416	102	176	212	232	227	219	125	93

(continued)

Table 5.3 (continued)

Number	Name	X-axis	Y-axis	0	1	2	3	4	5	6	7
				Xi'an	He yang	Ruyi	Longting	Zhichuan	Hancheng	Weinan	Huaxian
11	Tongchuan	410	414	124	180	194	214	209	223	130	150
12	Yijun	430	475	167	193	260	280	275	272	196	216
13	Huanglin	445	498	203	275	288	309	304	246	225	245
14	Luochuan	461	520	243	304	317	210	187	176	254	274

8	9	10	11	12	13	14
Huayin	Tongguan	Yaoxian	Tongchuan	Yijun	Huangling	Luochuan
0						
38	0					
65	27	0				
182	211	22	0			
248	277	65	43	0		
277	306	101	79	36	0	
306	335	141	119	76	40	0

Source The Author

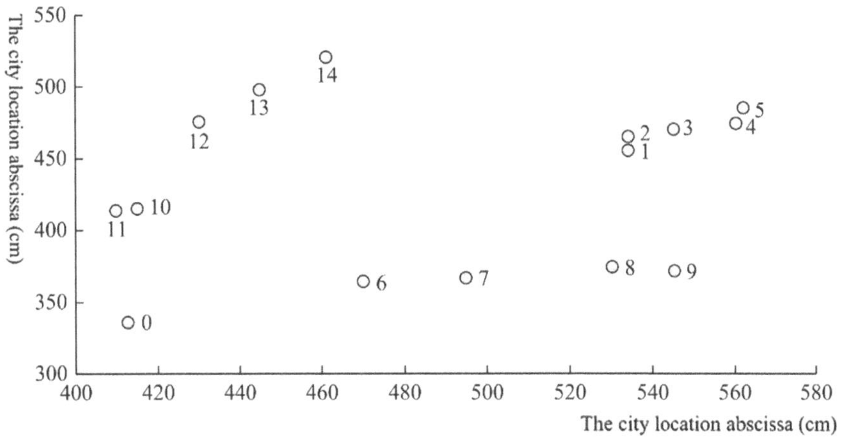

Fig. 5.3 Distribution of checkpoints along the route. *Source* The Author

energy expended during the postal vehicle's journey. The formula is:

$$\text{Carbon emission} = \text{carbon emission coefficient} \times \text{fuel consumption},$$

Fuel consumption is determined by factors such as transportation distance and cargo capacity (Sun & Yang, 2014); specifically, fuel consumption per unit distance (P) can be expressed as a linear function that is affected by postal load capacity (X) (Zhu & Li, 2013). Suppose the maximum load capacity of the vehicle is Q, then the fuel consumption per unit distance is P^* when the express delivery vehicle is fully loaded, and when empty it is P_0. And a and b are coefficients.

$$P_0 = \mathrm{a}Q_0 + \mathrm{b} \tag{5.4}$$

$$P^* = \mathrm{a}(Q_0 + \mathrm{X}) + \mathrm{b} \tag{5.5}$$

Combining the above two, we get

$$\mathrm{a} = \frac{P^* - P_0}{\mathrm{Q}} \tag{5.6}$$

Hence, the fuel consumption per unit distance $P(X)$ can be expressed as

$$\mathrm{P}(X) = p_0 + \frac{P^* - P_0}{\mathrm{Q}} X \tag{5.7}$$

Table 5.4 Parameter setting values

Parameters	Parameter value
Fuel consumption per unit distance when fully loaded	0.377 L/km
Fuel consumption per unit distance when empty	0.165 L/km
CO_2 emission coefficient e_0	2.63 kg/L
carbon tax C_0	0.0066 g/kg·km

Source The Author

The carbon emissions resulting from transporting mail from location *i* to location *j* for (*Qij*) are as follows

$$E = e_0 P\left(Q_{ij}\right)d_{ij} \tag{5.8}$$

Among them, e_0 is the CO_2 emission coefficient. The cost of carbon emissions can be calculated by multiplying the carbon tax (C_0) by the amount of carbon emissions, i.e. Carbon emission cost = carbon tax × carbon emission amount. Set carbon tax as C_0, then the carbon emission cost is

$$\mathrm{C} = c_0 \sum_{k=1}^{k} \sum_{i;j=0}^{n} x_{ij}^{k} e_0 P\left(Q_{ij}\right)d_{ij} \tag{5.9}$$

Parameter setting values are presented in Table 5.4.

5.2.2.2 Resolution Procedures of Ant Colony Optimization

1. Initialize Parameters

Firstly, this algorithm initializes relevant parameters, setting the number of ants as *m*, pheromone volatilization factor ρ, pheromone importance factor α, heuristic importance factor β, total pheromone release Q, initial value of iteration times *iter* = 1, maximum iteration number *iter_max*.

2. Construct the Solution Space

For each ant k (where k = 1, 2, …, m), a different starting point is chosen at random, and the subsequent generation visits cities in accordance with Eq. (5.3) until every city has been visited.

3. Update Pheromone

The first step is to calculate the path length of each ant L_k(k = 1, 2, …, m) and record the current optimal solution. Subsequently, the pheromone concentration between cities will be updated according to Formulae (5.4) and (5.5).

4. Assess the Need for Terminating

If the current number of iterations is less than the maximum number of iterations (i.e. $iter < iter\_max$), then $iter = iter + 1$. This implies that the path of each ant currently recorded should be cleared and the loop should return to step 2; otherwise, the loop should be terminated and the optimal solution should be output.

5.2.2.3 Simulation Results of Ant Colony Optimization

Table 5.3 shows that Matlab 7.0 was used to optimize and simulate postal routes 1, 2 and 3. Figure 5.4 indicates that after 100 iterations, ant colony optimization tends to be stable, suggesting that the shortest distance of the postal route at this point is close to the optimum.

As can be seen from Fig. 5.5, the optimal solution of ant colony optimization is to integrate the three existing postal routes into a single new route. The specified postal train dispatching route is:

$$Mail\ Route\ 1: 0–6–7–8–9–1–2–3–4–5–14–13–12–10–11–0.$$

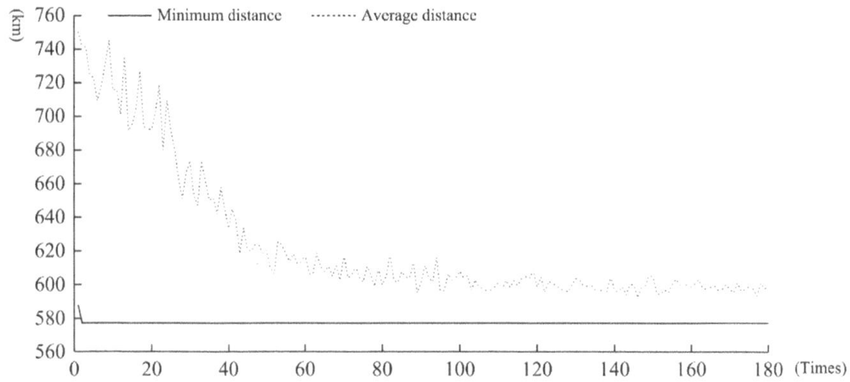

Fig. 5.4 The optimization process of ACO. *Source* The Author

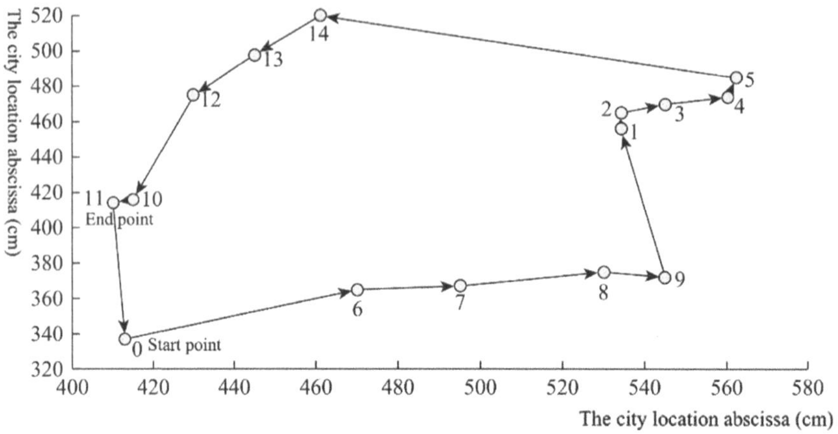

Fig. 5.5 Postal route optimized by ACO. *Source* The Author

That is: Xi'an—Weinan—Huaxian—Huayin—Tongguan—Heyang—Ruyi—Longting—Zhichuan—Hancheng—Luochuan—Huangling—Yijun—Yaozhou—Tongchuan—Xi'an.

The total distance of the postal route is 577.17 km, resulting in an emission of 377.70 kg of carbon.

The results of the experiment indicated that Ant Colony Optimization could effectively address the optimization problem of an express delivery network, resulting in a 794.83 km decrease in the optimized postal route distance.

5.3 Optimizing Express Delivery Networks with Simulated Annealing Algorithm

5.3.1 Simulated Annealing Algorithm

Simulated Annealing Algorithms (SAAs) are suitable heuristics for finding optimal solutions in global scope with high expected values. The algorithm was initially proposed by Kirkpatrick et al. who believed it to be analogous to the physical annealing process, which is a continuous iterative optimization. Thus, they designed SAAs for combinatorial optimization problems, which are based on the physical annealing process and a Monte Carlo iterative strategy (Kirkpatrick et al., 1983).

The physical annealing process of the Simulated Annealing Algorithm consists of three parts:

1 Heating Process

Its purpose is to continuously increase the temperature of the object. At high temperatures, the material is in a disordered state, displaying the solid thermal motion of particles. During the heating process, particles are made to deviate from their equilibrium position and their thermal motion is increased. When the temperature is high enough, the solid is converted into a liquid, thus promoting the transformation of the original non-uniform state of the system into a homogeneous form.

2 Isothermal Process

In the calculation process, a closed system exchanges heat with environment and remains at a constant temperature. In the iterative process, there is a spontaneous change in the system, and the energy level always decreases. When the energy reaches its minimum value, the system is in equilibrium.

3 Cooling Process

This process is designed to reduce the thermal motion of the particles, thus decreasing the energy and forming a crystal structure. It consists of two aspects: annealing, whereby the temperature of the object is gradually lowered by slow cooling until reaching a balanced and ordered state, i.e. the lowest energy state; and quenching, whereby the cooling rate is too rapid to enable energy minimization.

Among them, the heating process corresponds to the initial temperature setting of the algorithm, the isothermal process adopts the Metropolis criterion-based balance system (Li, 2008), and the cooling process is aimed at adjusting and controlling the stable reduction of parameters. The algorithm's fitness function is the change in energy, and the optimal solution is the lowest energy state value. The Metropolis criterion is the core in solving for the global optimum and is allowed to accept inferior solutions to a certain extent in order to avoid the algorithm from falling into a local optimum.

5.3.1.1 Mathematical Model of Simulated Annealing Algorithm

The energy of the particles can be used to represent the state of the material, and a mathematical model constructed with the aid of the Metropolis Algorithm is as follows:

1. Let the energy of the material in state i be $E(i)$. Affected by the external temperature T, the transition from state i to state j follows the following rules:

   If $E(j) \ll E(i)$, the accepting state is transitioned.
   If $E(j) > E(i)$, the state transition is accepted with the following probability.

$$e^{\frac{E(i)-E(j)}{KT}} \tag{5.10}$$

In which K is the Boltzmann constant and T is the temperature that the material accepts.

2. After numerous iterations, the material reaches thermal equilibrium at a certain temperature following sufficient energy conversion. At this time, the likelihood that the material in state i satisfies the Boltzmann distribution:

$$\mathrm{P}_T(x=i)=\frac{e^{\frac{E(i)}{KT}}}{\sum_{j\in s} e^{\frac{E(j)}{KT}}} \tag{5.11}$$

Among them, X is the random variable representing the material in its current state, and S is the state space set.

$$\lim_{T\to\infty}\frac{e^{\frac{E(i)}{KT}}}{\sum_{j\in s} e^{\frac{E(j)}{KT}}}=\frac{1}{|S|} \tag{5.12}$$

Among them, |S| is the number of the space state set S, indicates that, at high temperatures, all states have the same probability. However, when the temperature decreases, then:

$$\begin{aligned}\lim_{T\to 0}\frac{e^{\frac{E(i)-Emin}{KT}}}{\sum_{j\in s} e^{\frac{E(i)-Emin}{KT}}}&=\lim_{T\to 0}\frac{e^{\frac{E(i)-Emin}{KT}}}{\sum_{j\in S_{min}} e^{\frac{E(j)-Emin}{KT}}+\sum_{j\in S_{min}} e^{\frac{E(j)-Emin}{KT}}}\\ &=\lim_{T\to 0}\frac{e^{\frac{E(i)-Emin}{KT}}}{\sum_{j\in S_{min}} e^{\frac{E(i)-Emin}{KT}}}=\begin{cases}\frac{1}{|S_{min}|} & if\ i\in S_{min}\\ 0 & \text{others}\end{cases}\end{aligned} \tag{5.13}$$

Among them, $E_{min}=\min\limits_{j\in S} E(j)$ and $S_{min}=\{i|E(i)=E_{min}\}$.

Analysis of the formula (5.13) reveals that when the temperature is lowered to a shallow level, the material is likely to enter its minimum energy state, effectively solving the optimization problem of finding the minimum value.

5.3.1.2 Flow Chart of Simulated Annealing Algorithm

Figure 5.6 shows the flow chart of Simulated Annealing Algorithm.

5.3.2 *Analysis of Simulation Results*

5.3.2.1 Problem Description

It is assumed that all points along the postal routes are at the same elevation, with no impediments between them. As Table 5.2 illustrates, three postal routes on the automobile trunk line in Shaanxi Province were chosen as examples; Xi'an-Tongguan, Xi'an-Luochuan, and Xi'an-Hancheng and the Simulated Annealing Algorithm was employed to optimize the postal routes.

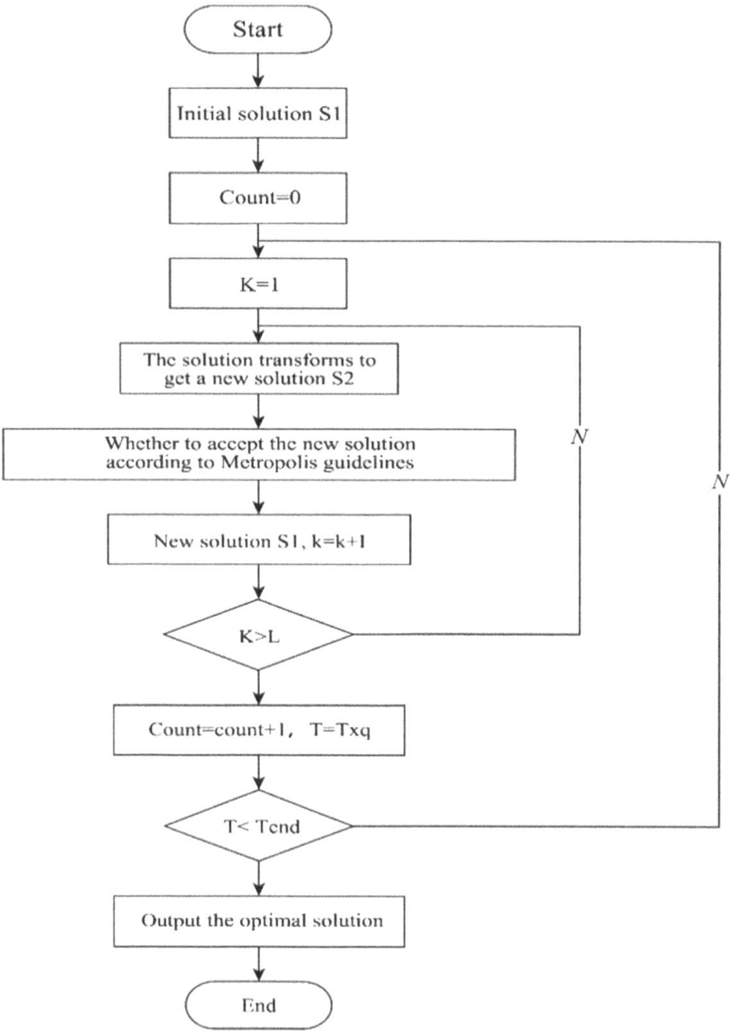

Fig. 5.6 Flow chart of simulated annealing algorithm. *Source* The Author

5.3.2.2 Resolution Procedures of Simulated Annealing Algorithm

1. Initialize Parameters

It is assumed that the initial temperature, T_0, is large enough, and that T is set equal to T_0. Any initial solution can be taken as S_1, and the number of iterations at each temperature, i.e., the Markov chain length, L, can be formulated.

2. Iterative Search

For the current temperature T and $K = 1,2,\ldots,L$, L times of tentative searches can be done according to the following procedure.

3. Perturbation Generates New Solutions

According to the nature of S_1, the current solution, a new solution S_2 is generated by random disturbance. The increment of S_2 is $df = f(S_2) - f(S_1)$, in which $f(S_1)$ represents the cost function of S_1. The smaller the value, the better.

4. Acceptance or Rejection of the New Solution According to Metropolis Guidelines

If $df < 0$, then we accept S_2 as the current solution, i.e. $S_1 = S_2$; otherwise, we calculate S_2 acceptance probability of $exp\left(-\frac{df}{T}\right)$, or a uniformly distributed random number *rand* in the (0,1) interval is randomly generated; if $exp\left(-\frac{df}{T}\right) > rand$, we also accept S_2 as the new current solution, i.e. $S_1 = S_2$, otherwise we keep the initial solution S_1.

5. Determining whether to Continue Iterating

When a new solution S_1 is generated, $k = k+1$. *If* $k < L$, then the iteration at this temperature has not ended. When one Mapkob chain length iteration has not been completed, the loop goes to step 3; if $k > L$, the number of iterations is incremented by 1.

6. Cooling down

When the temperature is less than the termination temperature, it will continue to cool down. Let $T = T \times q$, where q is the cooling factor, and $q < 1$.

7. Determining Whether to Terminate

It needs to determine if the termination temperature is reached; if not, the loop goes to step 2; if so, the optimal solution is output.

5.3.2.3 Simulation Results of Simulated Annealing Algorithm

According to Table 5.3, simulations of postal routes 1, 2 and 3 were optimized using Matlab 7.0. Figure 5.7 reveals that after 40 iterations the simulated annealing algorithm tends to stabilize, suggesting that the shortest distance of the postal route is close to the optimum.

As depicted in Fig. 5.8, the optimal solution of the simulated annealing algorithm is to combine the existing three postal routes into a single route. The specific postal train dispatch route is:

$$Mail\ Route\,1:\ 0\text{–}6\text{–}7\text{–}8\text{–}9\text{–}1\text{–}2\text{–}3\text{–}4\text{–}5\text{–}14\text{–}13\text{–}12\text{–}10\text{–}11\text{–}0.$$

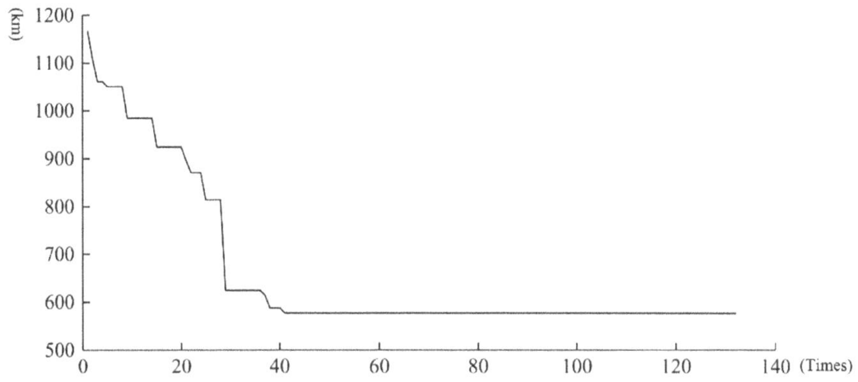

Fig. 5.7 The optimization process of simulation annealing algorithm. *Source* The Author

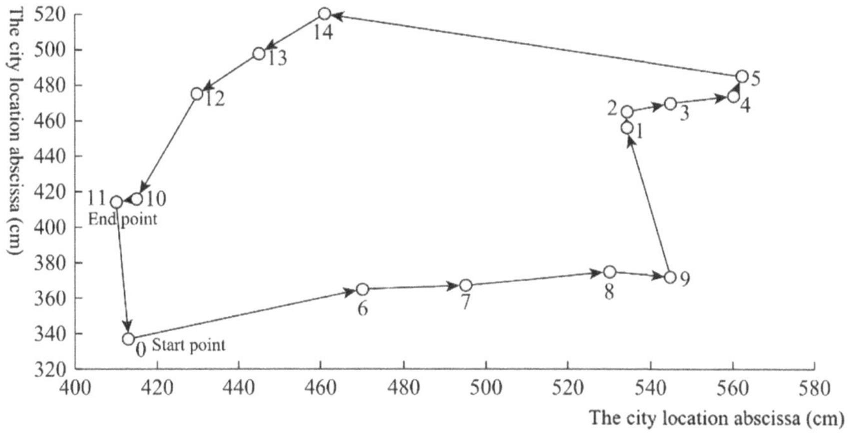

Fig. 5.8 Postal route optimized by Simulated Annealing Algorithm. *Source* The Author

That is Xi'an—Weinan—Huaxian—Huayin—Tongguan—Heyang—Ruyi—Longting—Zhichuan—Hancheng—Luochuan—Huangling—Yijun—Yaozhou—Tongchuan—Xi'an.

The total mileage of the postal route is 577.17 km with a total carbon emission of 377.70 kg.

The experimental results indicate that the simulated annealing algorithm is successful in tackling the optimization problem of express delivery network, resulting in a decrease of 794.83 km in the optimized postal route distance.

5.4 Summary

In this chapter, several common algorithms used in express delivery network optimization are introduced. By sorting out the content and models of the algorithms, it is evident that the performance characteristics of different algorithms in network optimization differ. The ant colony optimization and the simulated annealing algorithm are employed for express delivery network optimization and modeling. Three postal datasets of automobile trunk lines in Shaanxi Province are selected to simulate the two algorithms. The results demonstrate that both algorithms can effectively solve the problem of express delivery network optimization.

References

Amagai, Y., Shoji, K., & Toyama, F. (1983). Optimization by simmulated annealing. *Science, 42*(3), 671–680.

Dasgupta, D., & Michalewicz, Z. (1997). *Evolutionary algorithms in engineering applications. DBLP.*

Dorigo, M., Birattari, M. & Stutzle, T. (2006). Ant colony optimization. *Computational Intelligence Magazine IEEE, 1*(4), 28–39.

ErnÝ, V. (1985). Thermodynamical approach to the traveling salesman problem: An efficient simulation algorithm. *Journal of Optimization Theory and Applications, 45*(1), 41–51.

Holland, J. H. (1975). *Adaptation in Natural and Artificial Systems.* Ann Arbor: MI University of Michigan Press.

Holland, J. H. (1992). Genetic algorithms. *Scientific Ameracan*, 267, 66–72.

Huang, S. Y. (2010). Interpret the management specifications of postal dispatch. *China Post News* (2010, February06) 008.

Kennedy,J. & Eberhart,R. (1995). *Particle Swarm Optimization. Icnn95-international Conference on Neural Networks* [Conference]. IEEE, New York, USA.

Kirkpatrick, S., Gelatt, C. D. & Vecchi, M. P.(1983). Optimization by simulated annealing. *Science*, 220, 671–689.

Koza, J. R. (1992). *Genetic programming: on the programming of computers by means of natural selection.* Cambridge: MIT Press.

Li, P. (2008). New algorithm based on simulate anneal and hopfield neural for solving TSP. *Science Technology and Engineering, 8*(14), 3937–3939.

Rechenberg I. (1978). *Evolutionsstrategien. Simulationsmethoden in der Medizin und Biologie.* New York: Springer Berlin Heidelberg.

Song, H. B. & Song, Y. L. (2003). An improved genetic algorithm of vehicle routing problem with time window constrains. *Journal rkof Transport Information and Safety, 04*, 25–27.

Sun, K. & Yang, W. M. (2014). Research on dairy transportation and distribution in low-carbon economy. *Jiangsu Agricultural Sciences, 42*(4), 392–395.

Webster, B. & Bernhard, P.J. (2003). *A Local Search Optimization Algorithm Based on Natural Principles of Gravitation. Proceedings of the International Conference on Information and Knowledge Engineering* [Conference]. DBLP, Las Vegas, Nevada, USA.

Zhang, Z. X. & Shao, B. L. (2008). Vehicle routing and scheduling problems based on improved ACA. *Journal of Highway and Transportation Research and Development, 04*, 137–140.

Zhu, C. Z. & Li, Y. L. (2013). Research on vehicle path optimization with the smallest carbon emissions. *Computer Engineering and Applications, 49*(22), 15–18.

Chapter 6
Evaluation of the Development of Regional Express Delivery Networks

The express delivery industry in China started late but has grown significantly over the years. As the express delivery industry has become increasingly entrenched in people's lives, it has grown rapidly in our country. Nevertheless, issues with the express delivery industry, including an inadequate express delivery network structure, inadequate operations, and an uneven growth of regional express delivery, are hindering the industry's development. Consequently, factor analysis is used in this chapter to summarize the key factors affecting the development of express delivery networks and the distribution of regional express delivery development levels through empirical analysis.

6.1 Development of Regional Express Delivery Networks

The development level of the regional express delivery network is closely related to that of the regional express delivery industry. The development of the regional express delivery networks will drive the advancement of the regional express delivery industry, and if its development level is low, so is that of the regional express delivery industry. Therefore, it can be concluded that the development level of the regional express delivery networks is equal to that of the regional express delivery industry. Furthermore, for the sake of convenience in data collection and analysis, it is agreed that the development level of the express delivery industry in each province represents the development level of the regional express delivery networks.

Express delivery network is a unified system, organized in accordance with specific principles and operating guidelines to provide express delivery. All components of the system are interdependent and rely upon the overall efficiency of the network to fulfill the delivery. The primary entities of the national express delivery industry comprise of three parts: international express delivery enterprises , state-owned express delivery enterprises, and private express delivery enterprises.

P. Li and J. Wu, *Research on Innovation of Express Delivery Network Management*,
https://doi.org/10.1007/978-981-99-8982-9_6

The advantage of international express delivery enterprises is their mature international express delivery market and global network of freight capabilities, with large service providers such as DHL, FedEx, TNT and UPS. State-owned express delivery enterprises are characterized by their sound domestic express outlets and transport routes, such as Postal Express and China Railway Express Co. Private express delivery enterprises, in contrast, primarily focus on regional express agencies and have some influence.

Express delivery network serves as the connection between customers and express delivery enterprises, and delivery efficiency and punctuality are critical measures of the quality of express services. Following years of expansion, the leading international express delivery enterprises have progressively improved due to the continuous optimization and adjustment of the express delivery network globally. The network structures and functions of most China's express delivery enterprises are historically unreasonable and imperfect, necessitating an urgent need for improvement in regards to operational efficiency and overall service level. Moreover, due to insufficient implementation of the economic system reform and inadequate guidance and regulation of the market by the government, the development of express delivery networks has been complicated. Thus, it is of paramount importance to evaluate the state of China's express delivery networks.

6.2 Effective Evaluation Methods

The development of express delivery networks is at a similar level to that of the regional express delivery industry. The level of development of regional express delivery networks can be compared to the express delivery industry in each province in order to assess the development of the regional express delivery industry. Furthermore, it is necessary to evaluate the development level of the regional express delivery networks, which can be done through the use of principal component analysis, structural equation models, data envelopment analysis, and factor analysis.

6.2.1 Introduction of Evaluation Methods

6.2.1.1 Principal Component Analysis

Principal Component Analysis (PCA) plays an important role in multivariate statistics, as it transforms the original variable space of high dimensions into a new synthetic variable space of low dimensions. This reduces the complexity of the problem and simplifies data analysis by reducing the data dimension (Cai et al., 2008). The variables in the low-dimensional synthetic variable space are typically referred to as principal components, and have special relationships with the original variables in the high-dimensional space: (1) The principal components can be obtained by taking a linear combination of the original variables; (2) The dimension of the principal component space is much smaller than that of the original variable space when considered from a dimensional standpoint; (3) For the mapping implemented via mathematical transformation, the elements that have little to no relevance to the data are typically discarded, with only the main component (principal component) being chosen in order to retain the main information of the data without needing a large quantity of data like that of the original variable space; (4) From the perspective of the correlation of variables within the space, the principal component is uncorrelated, while the original variables are often correlated, leading to a high degree of redundancy.

(1) Brief Introduction of the Model

In the analysis of a statistical problem, we suppose that n samples are collected, each observed with p variables (denoted as $x_1, x_2, \ldots x_p$ with x_i having a mean value of 0, variance of 1, and $1 \leq i \leq p$) forming an $n \times p$ array of sample data X.

$$X = \begin{bmatrix} x_{11} \; x_{12} & \ldots & x_{1p} \\ x_{21} \; x_{22} & & x_{2p} \\ \vdots & \ldots & \vdots \\ x_{n1} \; x_{n2} & \cdots & x_{np} \end{bmatrix}. \tag{6.1}$$

Principal Component Analysis (PCA) utilizes p original variables $(x_1, x_2, \ldots x_p)$ to construct a few uncorrelated synthetic variables as a linear combination of the original variables. The new variables contain a majority of the information from p original variables. If we set $x_{1,} x_2, \ldots x_p x_1$ as original variables and $z_1, z_2, \ldots z_p (m < p)$ as new synthetic variable indicators, each of the synthetic variables is a linear combination of p original variables.

$$\begin{cases} z_1 = l_{11}x_1 + l_{12}x_2 + \cdots + l_{1p}x_p \\ z_2 = l_{21}x_1 + l_{22}x_2 + \cdots + l_{2p}x_p \\ \vdots \\ z_{\mathrm{m}} = l_{m1}x_1 + l_{m2}x_2 + \cdots + l_{mp}x_p \end{cases} \tag{6.2}$$

The following conditions are required simultaneously:

i. z_i and z_j are independent of each other;
ii. z_1 is the largest variance of all linear combinations of x_1, x_2, …, x_p;
iii. z_2 is the largest variance of all linear combinations of z_1 with uncorrelated x_1, x_2, …, x_p; z_m is the largest variance of all linear combinations of z_1, $z_2, \ldots z_{m-1}$ with uncorrelated $x_1, x_2, \cdots, x_p$, respectively. The new variables z_1, $z_2, \ldots z_m$ are called the first, second, …, and the mth principal components of x_1, x_2, …, x_p, respectively.

(2) Steps of Algorithm Implementation

i. Collecting data to get the raw sample data matrix

$$X_{n\times p} = \begin{bmatrix} x_{11} & x_{12} & \cdots & x_{1p} \\ x_{21} & x_{22} & & x_{2p} \\ \vdots & & \cdots & \vdots \\ x_{n1} & x_{n2} & \cdots & x_{np} \end{bmatrix} \quad (6.3)$$

ii. Standardizing Raw Data

Let $x_{ij}^{*} = \left(x_{ij} - \overline{x_j}\right)/\sqrt{s_j}$, where $\overline{x_j}$, s_j are the sample mean value and variance of the elements in column j respectively and $\overline{x_j} = \frac{1}{n}\sum_{k=1}^{n} x_{kj}$, $s_j = \frac{1}{n}\sum_{k=1}^{n}\left(x_{kj} - \overline{x_j}\right)^2$, then $X^* = \left(x_{ij}^{*}\right)_{n\times p}$ is the normalized number matrix.

iii. Calculating the covariance matrix (the method of directly calculating the correlation matrix of the raw data is employed in this chapter). The equation is as follows:

$$R = \begin{bmatrix} r_{11} & r_{12} & \cdots & r_{1p} \\ r_{21} & r_{22} & & r_{2p} \\ \vdots & & \cdots & \vdots \\ r_{p1} & r_{p2} & \cdots & r_{pp} \end{bmatrix} \quad (6.4)$$

Among which, $r_{ij}(i, j = 1, 2 \cdots p)$ is the correlation coefficient between the original variable x_i and x_j and r_{ij} and r_{ji}. The calculation formula is as follows:

$$r_{ij} = \frac{\sum_{k=1}^{n}\left(x_{ki} - \overline{x_i}\right)\left(x_{kj} - \overline{x_j}\right)}{\sqrt{\sum_{k=1}^{n}\left(x_{ki} - \overline{x_i}\right)^2 \sum_{k=1}^{n} x_{ki} - \overline{x_j}\right)^2}} \quad (6.5)$$

iv. Calculating the characteristic root and the corresponding standard orthogonal vector

v. Calculating the variable contribution rate g_i and the cumulative contribution rate h_i (i = 1, 2, …, p) in the space of low-dimensional synthetic variables

$$g_i = \lambda_i / \sum_{k=1}^{p} \lambda_k. \quad (6.6)$$

$$h_i = \sum_{k=1}^{i} \lambda_k / \sum_{k=1}^{P} \lambda^k \quad (6.7)$$

vi. Determining the number of retained principal components, and usually selecting the principal components with a cumulative contribution rate of 85–95% (Feng, 2007);
vii. Analyzing, Interpreting, Inferencing.

According to the cumulative contribution rate calculated by the software, components with a contribution rate of between 85 and 95% will be selected. Subsequently, an analysis will be conducted to determine the indicators that each component is composed of, their meaning will be explained, and the main indicators that influence the evaluation will be inferred. Finally, relevant suggestions and comments will be given.

6.2.1.2 Structural Equation Modeling

Structural Equation Modelling (SEM) is a statistical technique used to analyse the interrelationships between variables based on the covariance matrix of said variables. This technique involves both observed and latent variables, which are those that cannot be directly observed. Furthermore, SEM can be used to clearly assess the effect of individual indicators on the construct as a whole and the relationship between individual indicators (Wang & Cheng, 2012). An example of the SEM schematic diagram is depicted in Fig. 6.1.

(1) Brief Introduction of the Model

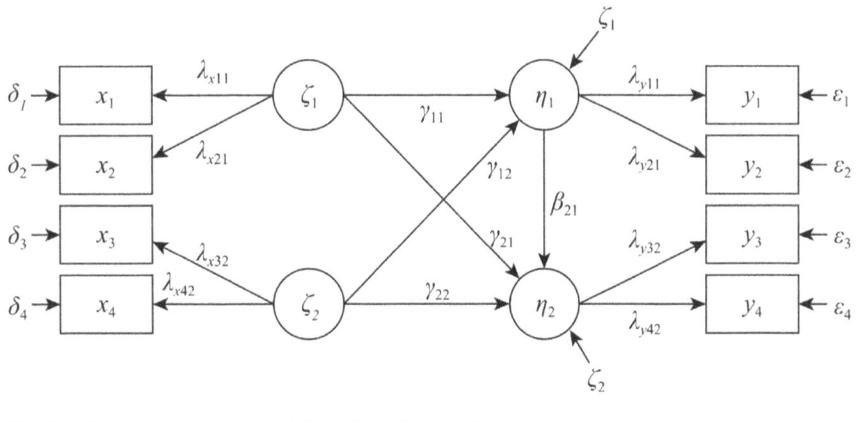

Fig. 6.1 Structural equation modeling. *Note* Observed variables are represented as rectangles, with exogenous observed variables denoted by and endogenous observed variables denoted by. Latent variables are usually represented as ellipses or circles, with exogenous latent variables denoted by ξ and endogenous latent variables denoted by η. Additionally, δ denotes the error associated with the exogenous observed variable, and ε denotes the error associated with the endogenous observed variable (*Source* The Author)

The modelling comprises of two components: a measurement equation and a structural equation. The measurement equation explains the relationship between latent variables and indicators, while the structural equation outlines the relationship between latent variables.

i. The measurement equation can be expressed as:

$$\begin{cases} x = \Lambda_x \eta + \delta \\ y = \Lambda_y \eta + \varepsilon \end{cases}. \tag{6.8}$$

Λ_x: Factor load matrix of exogenous observed variables on exogenous latent variables;

Λ_y: Factor load matrix of endogenous observed variables on endogenous latent variables;

δ: The error of exogenous observed variable x;

ε: The error of the endogenous observed variable.

ii. The structural equation can be expressed as:

$$\eta = B\eta + \Gamma\xi + \zeta \tag{6.9}$$

B—the path coefficient; Γ—the path coefficient;

ζ—the residual term of the structural equation which reflects the portion of the equation that is not explained.

(2) Steps of Algorithm Implementation

i. modeling construction

   The relationship between observed variables (indicators) and latent variables (factors), as well as the relationship between latent variables, are initially determined.

ii. modeling fitting

   Modeling parameters are estimated through a process known as modeling calculation.

iii. Modeling evaluation

   a. The appropriateness of the solution for the structural equation is assessed;
   b. The parameters are in agreement with the assumptions used in the modeling.;
   c. It is evaluated whether each goodness-of-fit index meets the requirements of the model;
   d. The model's fit to the data (NFI, NNFI, CFI, IFI, GFI, AGFI, RFI greater than 0.9, RMR less than 0.035, and RMSEA value less than 0.08) is tested, indicating that the model and data fit well.

iv. Modeling correction
   a. A priori modeling based on the relevant assumptions is proposed;
   b. The relationship between latent variables and indicators is tested, and a measurement modeling is established;
   c. The standard errors, t-values, standardized residuals, correction indices and various fit indices of each modeling are tested, and the modeling is modified accordingly. This process is then repeated;
   d. Another sample is tested.

6.2.1.3 Data Envelopment Analysis

Data Envelopment Analysis (DEA) is a quantitative method of evaluation that was introduced in the early 1980s to assess the relative effectiveness of comparable units of the same type based on multiple input and output indicators (Li, 2012).

(1) Brief Introduction of the Model

The Data Envelopment Analysis (DEA) methodology has two models: the Constant Returns to Scale (CCR) model, which was proposed by Charnes, Cooper, and Rhodes in the late 1970s, and the Variable Returns to Scale (BCC) model, which was proposed by Banker, Charnes and Cooper in the early 1980s. The former is based on the assumption of constant returns to scale, while the latter assumes variable returns to scale.

The CCR model can be expressed as:

$$\alpha^* = \min\left[\alpha - \varepsilon\left(\sum_{i=1}^{m} S_i^- + \sum_{l=1}^{m} S_l^+\right)\right]$$

$$s.t.\begin{cases} \sum_{j=1}^{n} x_{ij} w_j + S_i^- = \theta x_{ij} & i = 1, 2 \ldots m \\ \sum_{j=1}^{n} y_{lj} \lambda_j - S_l^+ = y_{lj} & l = 1, 2, \ldots s \\ \lambda_j, S_i^-, S_J^+ \geq 0 & j = 1, 2, \ldots n \end{cases} \tag{6.10}$$

Among which α^* denotes the comprehensive efficiency value derived by applying the CCR model; both S_i^- and S_l^+ denote the relaxation variable, S_i^- refers to output redundancy and S_l^+ refers to input redundancy; ε refers to the non-Archimedean infinitesimal.

The BCC model can be expressed as:

$$\beta^* = \min\left[\beta - \varepsilon\left(\sum_{i=1}^{m} S_i^- + \sum_{i=1}^{m} S_J^+\right)\right]$$

$$s.t.\begin{cases} \sum_{j=1}^{n} x_{ij} w_j + S_i^- = \theta x_{ij}, \; i = 1, 2 \ldots m \\ \sum_{j=1}^{n} y_{lj} \lambda_j - S_l^+ = y_{lj}, \quad l = 1, 2, \ldots s \\ \sum_{j=1}^{n} \lambda_j = 1 \\ \lambda_j, S_i^-, S_J^+ \geq 0 \qquad\qquad j = 1, 2, \ldots n \end{cases} . \tag{6.11}$$

Among which β^* denotes the pure technical efficiency value derived by using the BCC model; both S_i^- and S_l^+ denote the relaxation variable, S_i^- refers to output redundancy, and S_l^+ refers to input redundancy; ε refers to the non-Archimedean infinitesimal.

(2) Steps of Algorithm Implementation

i Identifying a single decision-making unit.;
ii Utilizing multiple decision-making units as evaluation groups and calculating the weight coefficient of input indicators based on the input-output ratio of the decision-making units.;
iii Assessing the efficacy of EDA through the observational data of each decision-making unit.

With n Decision-Making Units (DMUs), each denoted as $DMU_j (j = 1, 2, \cdots n)$ while the input volume x_{ij} of the ith category and output volume y_{lj} of the lth category are known in each of them. Furthermore, C w_j is the weight of the indicator, and λ_j is the input and output volume of the DMU, both of which are weighted.

In models (6.10) and (6.11), we assume that the optimal solution of the above model is θ^*, S^{-^*}, S^{+^*} The following conclusions can then be drawn:

If $\theta^* = 1$, it suggests that the associated decision-making unit (DMU) is consistent with weak efficiency data envelopment analysis (EDA).
If $\theta^* = 1$ and $S^{-^*} = S^{+^*} = 0$, it means that the corresponding decision-making unit DMU is valid for EDA;
If $\theta^* < 1$ or $S^{-^*} \neq 0$, $S^{+^*} \neq 0$, it means that the corresponding decision-making unit DMU is invalid for EDA.

6.2.1.4 Factor Analysis

In the process of data analysis, the amount of data is large and irregular, making it difficult to analyze directly. Factor analysis is used to simplify and regularize this data, as it initially appears to be erratic. The principle of factor analysis is to reduce the original variables into a few representative factors, which can capture most of the information in the original data, without causing too much loss to the original variables. Factor analysis uses a few factors to describe the relationship between indicators or factors, by summarizing multiple variables with close relationships into a certain category, and reflecting most of the information of the original data through a few factors.

In the analysis, the original variables need to be grouped according to the correlation between them. Moreover, there is a strong connection within the same group and a weak correlation across groups. One group of variables can be expressed as a structure, which can represent a certain aspect of the problem. By utilizing the contribution rate of variance as the weight to construct the evaluation function, factor analysis can effectively simplify the data (Yang et al., 2005). Simultaneously, most of the information of the original data is preserved, thus, factor analysis is objectively valid in analysis.

(1) Brief Introduction of the Model

Let p, the original varies $x_1, x_2, \ldots x_p$, have a mean of 0 and a standard deviation of 1. Each of the original variables is represented by a linear combination of f_1, $f_2, \ldots f_k$ (k < p), as shown in the Eq. (6.12).

$$\begin{cases} x_1 = a_{11} f_1 + a_{12} f_2 + a_{13} f_3 + \cdots + a_{1k} f_k + \varepsilon_1 \\ x_2 = a_{21} f_1 + a_{22} f_2 + a_{23} f_3 + \cdots + a_{2k} f_k + \varepsilon_2 \\ \qquad \cdots \\ x_p = a_{p1} f_1 + a_{p2} f_2 + a_{p3} f_3 + \cdots + a_{pk} f_k + \varepsilon_p \end{cases} \tag{6.12}$$

Equation (6.12) is an Output Model for factor analysis and can also be expressed as Matrix form:

$$X = AF + \varepsilon \tag{6.13}$$

Among these, F represents the factor. Since F appears in the linear expression of each original variable, it is also referred to as the common factor. is the factor load, that is, the loading of the ith variable on the jth factor. is a special factor to each corresponding variable and it is called as special factor. And the covariance of the X_i and F_j is:

$$\begin{aligned} Cov(x_i, f_j) &= Cov\left(\sum_{k=1}^{p} a_{ik} f_k + \varepsilon_i, f_j\right) \\ &= Cov\left(\sum_{k=1}^{p} a_{ik} f_k, f_j\right) + Cov(\varepsilon_i, f_j) = a_{ij} \end{aligned}$$

If is standardized, the standard deviation of is 1. Therefore, we have this equation:

$$r_{x_i, f_j} = \frac{Cov(x_i, f_j)}{\sqrt{D(x_i)}\sqrt{D(f_j)}} = Cov(x_i, f_j) = a_{ij} \tag{6.14}$$

For the standardized, is the correlation coefficient of and.and i.e. under the premise that the factors are not correlated, the factor load is the correlation coefficient of and, reflecting the degree of correlation between and. The larger the factor load, the closer the relationship between and and vice versa. At the same time, factor load reflects the importance and effects of on (Wen et al., 2014).

(2) Steps of Algorithm Implementation.

i Creating a data table.

A data table is created by using data from economic indicators for n regions in the *tth* year: $X^t = \left(x_{ij}^t\right)_{n\times p}$. Among these, $i = 1, 2, \ldots, n; j = 1, 2, \ldots, p$. Assuming the data to be analyzed has been derived from *T* years, *T* data tables can be established according to the time frame. For the sake of convenience in analysis, the T data tables can be combined by columns chronologically to form a single global data table:

$$X = \left(X^1, X^2, \cdots, X^T\right)'_{Tn\times p} = \left(x_{ij}^t\right)_{Tn\times p} \tag{6.15}$$

Among these, t = 1, 2,..., T. When considering only one time period, we set T = 1.

ii Data standardization

In order to avoid anomalies in data analysis due to the disparate dimensions of each indicator data during data collection, standardization of the data is employed using the following method

$$x_{ij} \Leftarrow \frac{x_{ij} - \overline{x}_j}{\sigma_j} \tag{6.16}$$

Among these, and are the mean and standard deviation of the *j*th indicator in each region. It is important to note that, for simplicity, is still used in the above equation to represent standardized data. Likewise, the global data table is still symbolized by *X*.

iii Calculation of Covariance Matrix

For the purpose of facilitating the calculation of the covariance matrix, the center of gravity of the *j*th indicator data is defined:

$$g_j = \sum_{i=1}^{n} q_{ij}\left(x_{ij} - \overline{x}_j\right) \tag{6.17}$$

Among these, indicates that the weighting coefficient used to sum each sample point, which obeys the equation: $\sum_{i=1}^{n} q_{ij} = 1$.

The *j*th indicator is extracted from the global data table X to form the global variable $X_j = \left(x_{1j}, \cdots, x_{nj}\right)$. Thus, the global variance of *Xj* can be defined as:

$$s_j = Var(X_j) = \sum_{i=1}^{n} q_{ij}(x_{ij} - \overline{x}_j)^2 \tag{6.18}$$

Thus, the global covariance,, can be obtained, among which, $c_{jk} = Cov(X_j, X_k) = \sum_{i=1}^{n} q_{ij}q_{ik}(x_{ij} - \overline{x}_j)(x_{ik} - \overline{x}_k)$

iv. Calculation of Eigenvectors

The eigenvalue of V is decomposed to obtain its p eigenvalues, which are then arranged in descending order as, with corresponding eigenvectors $\mu_1, \mu_{2,} ..., \mu_p$.

v. Determining the Principal Components

The contribution rate of variance associated with each eigenvector is calculated as follows:

$$s_j = Var(X_j) = \sum_{i=1}^{n} q_{ij}(x_{ij} - \overline{x}_j)^2 \tag{6.19}$$

Cumulative variance contribution rate is:

$$b_m = \sum_{k=1}^{m} a_k = \sum_{k=1}^{m} \lambda_k / \sum_{i=1}^{p} \lambda_i \tag{6.20}$$

It is common practice to select the first m feature vectors as the principal components when surpasses 85% for the first time to determine the principal components based on the cumulative variance contribution rate calculated (Liao et al., 2015).

vi. Calculation of Factor Load Matrix

The factor loading matrix is constructed among which represents the correlation coefficient of the variables and the factor, reflecting the degree of correlation between the two, $i = 1, 2, \ldots,$ m.

vii. Calculating the Principal Component Coefficients and Weights Corresponding to Each Index

The principal component coefficients of the indicators are specifically calculated as follows:

$$r_{X_{i,\mu_j}} = \frac{Cov(X_i, \mu_j)}{\sqrt{Var(X_i)\sqrt{Var(\mu_j)}}} = Cov(X_i, \mu_j) = a_{ij} \tag{6.21}$$

The weight of each indicator is:

$$\mathcal{W} = \sum_{\mathrm{i}=1}^{\mathrm{p}} \mathrm{a_{mi} a_i / p} \tag{6.22}$$

viii. Calculation of the Score Vector of Factors:

$$F_i = A'V^{-1}X_i \tag{6.23}$$

Among the above $X_i = (x_{i1}, \ldots x_{ip})'$.

ix. Calculation of the comprehensive score of each sample:

$$\mathrm{F} = \sum_{i=1}^{m} \lambda_i \mathrm{F_i} / \sum_{i=1}^{m} \lambda_i \tag{6.24}$$

where $\mathrm{F_i}$ is the denormalized ith principal component score.

6.2.2 *Comparison and Analysis of Evaluation Methods*

Several methods discussed in this chapter have advantages and disadvantages, as seen in Table 6.1.

The Principal Component Analysis (PCA), Structural Equation Modeling (SEM) and Data Envelopment Analysis (DEA) methods have been found to have limitations with regards to their application, as evidenced by the comparison of their benefits and drawbacks in the table. The Principal Component Analysis (PCA) is unique (Jia, 2009). The load coefficient matrix of Factor Analysis (FA) is not unique; a new factor matrix can be obtained through factor rotation. The FA method uses the variance contribution rate of each main factor, composed of the original variables, as the weight to construct the evaluation function, thus yielding an evaluation result that has a strong degree of objectivity. Although the Structural Equation Model (SEM) allows for errors in both the independent and dependent variables, it is not capable of performing nonlinear analysis, cannot handle qualitative data (Sun, 2005), and has stringent requirements for evaluation indicators, which must also be independent. The Data Envelopment Method (DEM) is a comprehensive technique used to evaluate the relative performance of entities within the same sector, considering multiple input and output indicators (Wei & Wang, 2011). The method is based on the premise of low correlation between indicators and a limited number of indicators. In this study, numerous indicators were used to assess the development level of the regional express delivery network, which differ in terms of dimension. The Factor Analysis Model is not affected by the dimensions of the indicators, rendering it a more effective tool for evaluating and analyzing the regional express delivery network development level than other methods. Therefore, Factor Analysis is employed in this book to evaluate and analyze the development level of the regional express delivery network.

Table 6.1 Comparison of advantages and disadvantages of evaluation methods

Evaluation method	Advantages	Disadvantages
Principal Component Analysis	It can simplify complex events through dimension reduction, and reflect most of the information during evaluation, which is scientific	When the load symbol of principal component factor is positive or negative, the comprehensive evaluation function has unclear meaning and low naming clarity. It is easy to make mistakes in the selection of principal components
Structural Equation Modeling	It can process multiple dependent variables at the same time, allow independent variables have the measurement errors, and estimate the fitting degree of the entire model, and on this basis, further allows for a more flexible measurement mode	If the required sample size is large and needs to meet the assumption of a multivariate normal distribution, which is rarely used for forecasting applications where the model is based on a real problem but may eventually produce unexplained results
Data Envelopment Analysis	When evaluating the efficiency of multiple inputs and multiple outputs with comprehensive indicators, it is not necessary to construct a production function to estimate the parameters, and the input-output dimension, measurement unit, and human factors will not affect its efficiency evaluation results	If the selected input and output items are inappropriate, it will affect the accuracy of the efficiency evaluation
Factor Analysis	The data can be simplified by regrouping the information from the original variables to find the common factors affecting the variables, providing a convincing basis for further analysis of the problem and making the factor variables easier to interpret through rotation	When calculating factor scores, the least squares method can be generally adopted, which has certain limitations and sometimes fails

Source The Author

6.3 Construction of Evaluation Index System of Regional Express Delivery Development

It is essential to accurately identify the various factors affecting the development of express delivery enterprises and to establish the evaluation indices for the competitive analysis of express delivery companies in order to ensure accurate and reasonable analysis and comparison of their competitiveness.

6.3.1 Principles of the Evaluation Index System

The selection and quantification of evaluation indices are the basis for the establishment of evaluation models, and are also the key to determining the quality of evaluation results. Consequently, it is essential to analyze the development of the regional express delivery industry scientifically and to establish an evaluation system. The selection of evaluation indices should adhere to these principles.

(1) Scientificity Principle.

In order to meet the standards of evaluation, the selection of evaluation indices should be scientifically-based and should meet actual needs while minimizing the influence of human-subjective factors.

(2) Purposiveness Principle.

In order to effectively evaluate the development of the express delivery sector, an evaluation index system should be established to accurately reflect the status of regional express delivery networks and the development of these networks, as well as any other pertinent factors. This index system should accurately reflect the status of express delivery and the development trends of express delivery companies in order to provide reasonable guidance for the development of express delivery companies.

(3) Availability Principle.

In order to ensure the feasibility of the evaluation, the selected indices should adhere to the availability principle, meaning they should be obtainable and be compatible with the current context.

(4) Practicality Principle.

To ensure the practicality of the evaluation of the development level of the express delivery network, the selection of evaluation indices should be easy to comprehend and enable the model to collect practical indicators, thus achieving operability of the evaluation, promotion and utilization of the effect, and resolution of practical problems.

6.3.2 Construction of the Evaluation Index System

Prior to evaluating the primary factors affecting the development level of the express delivery network, it is necessary to construct a corresponding evaluation index system for the development level of the regional express delivery industry. These factors span multiple fields and multiple levels of content, which are not only contingent on the development status of the express delivery industry itself, but also on the degree of development of the transportation network infrastructure, the influence of the macroeconomic development environment and the degree of informatization. Taking into account the development of national logistics and express delivery, the cross-sectional data of 31 provinces and municipalities (excluding Hong Kong, Macao and Taiwan) in 2016 were selected for factor analysis. These data were primarily sourced from the "National Bureau of Statistics of the People's Republic of China". The development level of express delivery network is evaluated across four aspects and 13 indicators, as indicated in Table 6.2.

i. The macroeconomic environment index measures regional Gross Domestic Product and residents' final consumption expenditure, for the purpose of reflecting the economic foundation for the development of logistics express delivery network in various regions.
ii. System inputs indices include the number of employees, the number of cars in operation, fixed asset investment of transportation, warehousing and postal industry, and the number of enterprises. The number of employees refers to those in the postal industry. These indicators reflect the investment in the logistics express delivery system from three perspectives: human, material and financial resources..
iii. The development index of the express delivery industry comprises railway mileage, road mileage, waterway mileage, freight volume, freight turnover, and the number of express delivery business outlets. These indicators reflect the business volume of the logistics express delivery industry and the development trend of the express delivery industry.
iv. The informatization degree index refers to the number of individuals who access the internet, which reflects the level of informatization in a given region.

Table 6.2 Evaluation index of express delivery network

Primary indicators	Secondary indicators
Macroeconomic environment	GDP x_1 (100 million yuan) Residents' final consumption expenditure x_2 (100 million yuan)
System input	Number of employees x_3 Car Ownership x_4 (10,000 vehicles) Fixed asset investment in transportation, warehousing and postal services x_5 (100 million yuan) Number of enterprises x_6 (pieces)
Development of express delivery industry	Mileage of railway transportation x_7. (KM) Mileage of road transportation x_8 (KM) Mileage of waterway transportation x_9 (KM) Freight volume x_{10} (Ten thousand tons) Volume of cargo turnover x_{11} (100 million ton-kilometers) Number of business outlets x_{12} (pieces)
Informatization level	The number of Internet users x_{13}

Source The Author

6.4 Evaluation of Regional Express Delivery Network Development through Factor Analysis

Common methods for evaluating the development level of regional express delivery have been outlined above. Each evaluation method has its own benefits and drawbacks and the empirical results are varied. In accordance with the requirements of the research purpose and content, the factor analysis method has been chosen to evaluate the development level of regional express delivery networks.

6.4.1 Instructions for the Implementation of Factor Analysis

Factor analysis generalizes and describes data by utilizing independent, interpretable and often unobservable factors to explain a set of interrelated variables. As these correlated factors are typically not directly observable, this type of analysis is usually undertaken with factor analysis.

6.4.1.1 Instructions for Controlling the Process of Factor Analysis

pro factor data = dataset <option list>;
priors common factor variance;
var variable table;
partial variable table;
freq variable;
weight variable;
by variable;

6.4.1.2 Option for Output Data Set

Out = output dataset. An output dataset is created, which mainly contains the entire input data set and the estimated score for each factor.

outstat = Output dataset. It is used to store the results of factor analysis. Stored factor results can also be used as read-in datasets for further factor analysis.

6.4.1.3 Options for Factor Extraction and Common Factor Variance

Method = factor selection methods. They are principal (principal component method), prinit (iterative principal factor method), usl (unweighted least squares factor method), alpha (factor method or Harris method), ml (maximum likelihood) method), image (principal component method of image covariance matrix), pattern (read factor model from the data set with type = option), score (read score coefficient from the data set with type = option). Commonly used methods are principal (principal component method), ml (maximum likelihood method) and prinit (iterative principal factor method).

Heywood Case. If the variance of the common factor is greater than 1, then the iteration can continue. Since the common factor variance is the square of the correlation coefficient, it is mathematically determined to be between 0 and 1. However, it is possible to exceed 1 in the iterative estimate of the common factor variance. If the common factor variance is equal to 1, this is referred to as the Heywood case; if the common factor variance is greater than 1, this is referred to as the super-Heywood case. In this instance, the factorial solution is not valid.

priors = the method of calculating the common factor variance. The method for calculating the prior common factor variance estimation is specified as assigning an initial value of 1.0 to the common factor variance of each variable, with "max" as the maximum absolute correlation coefficient, "smc" as the multiple correlation coefficient, "asmc" as proportional to the square of the multivariate correlation coefficient, but appropriately adjusted to make their sum equal to the maximum absolute correlation, "input" as a priori common factor variance estimate read from the data set specified by data= and type=, and "random" as a random number between 0 and 1.

6.4.1.4 Options for Specifying the Number of Factors and Convergence Criteria

factors = *n*—It requires keeping any common factors, but only those common factors with eigenvalues greater than 1 should be retained

Mineigen = *p*—It specifies the minimum eigenvalue of the retained factor

proportion = *p*—Using the prior common factor variance estimate, the proportion of the common variance for the retained factor is to be specified as this *p* value

converge = *p*—Stop iteration when the maximum change in the commensurate variance is less than *p*. And default = 0.001

maxiter = *n*—It specifies the maximum number of iterations. The default value is 30

6.4.1.5 Options for the Rotation Method

rotate = factor rotation mode name—It gives the rotation method including none, varimax, quartimax, equamax, orthomax, hk, promax, procrutes. Commonly used methods are varimax (orthogonal maximum variance axis method), orthomax (orthogonal variance maximum axis method specified by gamma = weight) and promax (oblique rotation based on orthogonal maximum variance axis).

norm = kaiser I raw I weight I cov I none. The arrangement of the rows in the model matrix can be defined to facilitate a smooth subsequent factor rotation. The Cureton-Mulaik method is used for weighting when the 'norm=weight' option is selected; when the 'norm=cov' option is selected, the rows of the model matrix are rescaled to represent covariance rather than correlation coefficients; and when the 'norm=raw or none' option is selected, no normalization is applied.

gamma = p—It specifies the maximum rotational weight of the orthogonal variance.

prerotate = factor reel method name—It specifies the method of prerotation. In addition to the rotation methods of promax and procrutes, any other rotation method can be used.

6.4.1.6 Options for Controlling the Printout

simple—It is to print output that includes basic statistics

corr—It is to print the correlation and partial correlation matrices

score—It is to print the coefficients in the factor score model

scree—It is to print a screen plot of the eigenvalues

ev—It is to print out eigenvector

residuals—It is to print residual correlation matrices and related partial correlation matrices

nplot = *n*—It specifies the number of factors to be plotted

plot—It plots the factor model after rotation

preplot—It plots the factor model before rotation

msa—It is to print the partial correlation between each pair of variables controlled by all remaining variables and to sample the appropriate Kaiser measure

reorder—This means that the rows of the various factor matrices are reorganized when printed out, with the variables having the highest absolute loadings on the first factor being output first, followed by the variables with the highest absolute loadings on the second factor, and so on

6.4.2 *Empirical Steps of Factor Analysis*

6.4.2.1 KMO Inspection

The KMO and Bartlett's test statistics (shown in Table 6.3) can be obtained by SPSS. The KMO test value of 0.793, which is greater than 0.7, indicates that the selected indicators have certain commonalities. The significance of the Bartlett sphericity test

Table 6.3 KMO and Bartlett's test statistics

The adequate sample	Kaiser-Meyer-Olkin be measured		0.793
Bartlett's sphericity test		Approximate chi-square	517.746
		df	78
		Sig	0.000

Source The Author

(P = 0.000) being less than 0.01 further indicates that the indicators are not correlated with each other, thus allowing for the following analysis to be completed.

6.4.2.2 Principal Component Extraction

After determining the validity of the global principal component, the data was standardized through the use of SPSS software, a process also known as nondimensionalization. The variance contribution rate, cumulative variance contribution rate, and the load matrix are then obtained, as shown in Table 6.4. It is generally accepted that the cumulative contribution rate should not be less than 85%, and the eigenvalue should not be less than 1. According to the results calculated by the software, three principal components have been selected, whose cumulative contribution rate has exceeded 85% and reached 86.211%. This is sufficient to demonstrate that the three principal components can accurately explain the information contained in all of the original indicators.

The results obtained by rotating the matrix with maximum variance are displayed in Table 6.5, which shows the loadings of each variable on the principal components. It can be seen from Table 6.5 that the five factors in GDP, final consumption expenditure, postal employment, the number of enterprises, and cargo turnover have relatively large load values in the first principal component, reflecting the social and macroeconomic conditions. Thus, it is named the economic base factor. The second principal component, representing investment in the logistics, express delivery industry and potential customers, is composed of fixed asset investment in transportation, warehousing and postal services, waterway transportation mileage, the number of express outlets and the number of Internet users. This is referred to as the input factor. The third principal component, representing the business development status of the logistics, express delivery industry, comprises the number of operating vehicles, railway transportation mileage, road transportation mileage and freight volumeand is referred to as an industrial development factor.

Table 6.4 Explained total variance

Elements	Initial eigenvalues			Extract sums of squared loadings			Rotation sums of squared loadings		
	Total	Variance contribution rate (%)	Cumulative contribution rate (%)	Total	Variance contribution rate (%)	Cumulative contribution rate (%)	Total	Variance contribution rate (%)	Cumulative contribution rate (%)
1	8.126	62.511	62.511	8.126	62.511	62.511	5.265	40.498	40.498
2	1.411	10.856	73.368	1.411	10.856	73.368	3.519	27.067	67.564
3	1.235	9.500	82.867	1.235	12.884	86.211	1.989	15.303	82.867
4	0.721	5.543	88.410						
5	0.487	3.744	92.154						
6	0.376	2.895	95.049						
7	0.231	1.777	96.826						
8	0.190	1.465	98.292						
9	0.117	0.897	99.189						
10	0.057	0.440	99.629						
11	0.030	0.233	99.862						
12	0.014	0.109	99.971						
13	0.004	0.029	100.000						

Source The Author

Table 6.5 Rotation component matrix

Index	Elements		
	1	2	3
Volume of cargo turnover (100 million ton-kilometers)	0.903	0.050	0.163
Number of employees in transportation, warehousing and postal services (10,000 people)	0.733	0.481	0.109
Final consumption expenditure	0.708	0.673	0.149
GDP (100 million yuan)	0.707	0.641	0.216
Number of enterprises	0.703	0.654	0.070
Mileage of waterway transportation (KM)	0.173	0.851	-0.027
Number of business outlets (pieces)	0.533	0.805	0.158
Fixed asset investment in transportation, warehousing and postal services (100 million yuan)	0.321	0.688	0.535
Number of Internet users (10,000 people)	0.575	0.681	0.371
Mileage of railway transportation (KM)	0.124	−0.167	0.843
Mileage of road transportation (KM)	−0.098	0.560	0.760
Number of vehicles in operation (10,000/unit)	0.508	0.223	0.718
Freight volume (10,000 tons)	0.562	0.280	0.569

Source The Author

6.4.2.3 Calculating Factor Score and Comprehensive Score of Each Province

The three principal components extracted are F1, F2, and F3. Table 6.6 presents the score coefficient matrix for each factor.

The calculation formulas for the three principal components can be found in Table 6.5:

$$\begin{aligned} \mathrm{F1} &= 0.119x_1 + 0.117x_2 + 0.206x_3 - 0.133x_6 + 0.445x_{11} \\ \mathrm{F2} &= 0.203x_5 + 0.402x_9 + 0.228_{12} + 0.124x_{13} \\ \mathrm{F3} &= 0.280x_4 + 0.439x_7 + 0.353x_8 + 0.188x_{10} \end{aligned}$$

By weighing the three principal components according to their respective eigenvalue contribution rates and summarizing them simultaneously, we can infer the current comprehensive score calculation formula for the development of the express delivery industry.

$$\mathrm{F} = (\mathrm{F1} \times 62.511 + \mathrm{F2} \times 10.856 + \mathrm{F3} \times 12.884)/86.211$$

Table 6.6 Component score coefficient matrix

Index	Elements		
	1	2	3
Final consumption expenditure (100 million yuan)	0.117	0.094	−0.091
GDP (100 million yuan)	0.119	0.071	−0.053
Number of enterprises (pieces)	0.133	0.094	−0.130
Number of employees in in transportation, warehousing and postal services (ten thousand people)	0.206	−0.015	−0.096
Fixed asset investment in transportation, warehousing and postal services (100 million yuan)	−0.151	0.203	0.163
Mileage of road transportation (KM)	−0.363	0.262	0.353
Mileage of railway transportation (KM)	0.008	−0.222	0.439
Mileage of waterway transportation (KM)	−0.210	0.402	−0.125
Number of Internet users (10,000 people)	0.012	0.124	0.043
Number of operating vehicles (10,000/unit)	0.091	−0.135	0.280
Freight volume (10,000 tons)	0.122	−0.109	0.188
Cargo turnover (100 million ton-kilometers)	0.445	−0.315	−0.048
Number of express outlets (pieces)	−0.026	0.228	−0.075

Source The Author

According to data from thirteen indicators in thirty-one regions, the measurement level of each principal component was obtained through factor analysis. The combined score formula (F = (F1*62.511 + 2*10.856 + F3*12.884)/86.211) was used to obtain scores for each region, and their respective rankings were shown in Table 6.7.

6.5 Analysis of Empirical Results

6.5.1 Role of Economic Conditions in Regional Logistics and Express Delivery Industry Development

The analysis of the above results shows that there are three principal components that measure the development of the logistics and express delivery industry through the validity of the model and data. These are referred to as economic basic factors, input factor factors, and industrial development factors. Among them, the economic basic factor is the first main factor, accounting for 62.511%; the second main factor, the input factor factor, is 10.856%; and the third main factor, the industrial development

Table 6.7 Scores and rankings of logistics and express delivery industry development of 31 provinces in China

Region	F1 scores	F2 scores	F3 scores	Overall scores	Rankings
Beijing	−2.17	2.04	1.32	0.32	10
Tianjin	−0.78	−0.53	0.10	−0.12	12
Hebei	2.06	−1.13	−0.87	1.00	5
Shanxi	0.19	−0.76	−0.58	−0.16	13
Neimenggu	−0.40	0.15	−0.39	−0.19	14
Liaoning	1.45	0.96	0.43	0.99	6
Jilin	0.93	−0.87	4.69	−0.48	21
Heilongjiang	1.27	−0.61	−0.58	−0.50	22
Shanghai	−0.29	2.70	0.36	1.41	2
Jiangsu	−0.08	0.50	−0.32	1.13	4
Zhejiang	0.05	0.05	−0.08	0.77	7
Anhui	2.38	1.94	0.04	0.65	8
Fujian	−0.21	−0.11	0.48	0.02	11
Jiangxi	−1.10	−0.69	−0.02	−0.36	18
Shangdong	0.43	0.69	−0.67	1.14	3
Henan	0.24	−0.15	0.90	0.63	9
Hubei	1.71	−1.04	−0.25	−0.21	15
Hunan	0.09	−0.47	−0.58	−0.36	19
Guangdong	−0.89	−0.98	−0.24	2.02	1
Guangxi	−0.68	−0.48	0.10	−0.25	16
Hainan	−1.13	−0.74	−0.01	−0.65	25
Chongqing	−0.20	0.11	−0.39	−0.42	20
Sichuan	−0.73	−0.12	−0.46	−0.61	23
Guizhou	0.18	2.14	−1.08	−0.83	28
Yunan	−0.21	0.17	−0.39	−0.86	29
Xizang	−0.32	−0.33	−0.56	−0.98	31
Shaanxi	−1.16	−0.84	−0.28	−0.25	17
Gansu	−0.36	−0.35	−0.10	−0.69	27
Qinghai	0.58	−0.48	−0.17	−0.89	30
Ningxia	−0.55	−0.44	−0.20	−0.67	26
Xinjiang	−0.35	−0.34	−0.15	−0.62	24

Source The Author

factor, is 12.884%. This indicates that the explanation degrees of each main factor to the development level of the logistics express delivery network are 62.511%, 10.856%, and 12.884% respectively. Consequently, the basic economic factor is the key factor for the development of the logistics and express delivery industry, the industrial development factor is the secondary factor, and the input factor factor is the third factor. These three factors jointly promote and largely determine the development level of the express delivery industry.

6.5.2 Regional Division and the Development of the Regional Express Delivery Industry

According to Table 6.7, the ranking of the development level of logistics and express delivery industry in 31 regions in China is summarized according to the division of administrative regions, and Table 6.8 is obtained. It can be seen that the development of regional logistics in China is related to the division of regions, which can be divided into three levels. The first tier ranks in the top 11, of which there are a total of nine provinces and cities located in the east. The second tier ranks from 12 to 19th, with four central regions and three provinces located in the west. Consequently, the second level can be labeled as the central region; the third level is ranked in the bottom 12, and a total of nine provinces belong to the western region, thus naming it the western region. To conclude, the development of the logistics and express delivery industry is related to the comprehensive development environment of the region, and the development of different regions has a distinct gap.

Table 6.8 Distribution of region by regional overall scores

Geographical division	Province and ranking
Eastern regions of China	Guangdong (1), Shanghai (2), Shandong (3), Jiangsu (4), Hebei (5), Liaoning (6), Zhejiang (7), Beijing (10), Fujian (11), Tianjin (12), Hainan (25)
Central regions of China	Anhui (8), Henan (9), Shanxi (13), Hubei (15), Jiangxi (18), Hunan (19), Jilin (21), Heilongjiang (22)
Western regions of China	Neimenggu (14), Guangxi (16), Shaanxi (17), Chongqing (20), Sichuan (23), Xinjiang (24), Ningxia (26), Gansu (27), Guizhou (28), Yunnan (29), Qinghai (30), Xizang (31)

Source The Author

6.5.3 *Regional Concentration Trend in Development*

The express delivery industry in the eastern provinces has achieved a high level of development overall. Guangdong, Shanghai, Shandong, Jiangsu, Hebei, Liaoning and Zhejiang provinces constitute the top seven in the ranking, while Beijing, Fujian and Tianjin provinces are among the top 12, with only Hainan Province ranking relatively low. The development level of the logistics and express delivery industry in the eastern region is concentrated, with the gradient of the development level gap between provinces being relatively slow, indicating the presence of a concentrated development trend with a clear scale effect. The overall development level of the regional industry is high, with the overall score being generally greater than zero. By contrast, the central and western regions exhibit a concentration of industrial development, albeit at a lower level. The overall score of cities and regions in the central and western regions are all negative, thus suggesting a large gap in comparison to the eastern regions.

6.6 Summary

This chapter introduces commonly-used evaluation methods for the development of regional express delivery networks. A factor analysis method was employed, under a comprehensive balance, to conduct empirical research to evaluate the development of national express delivery networks. By analyzing 13 indicators from four aspects, we obtained the principal factors that promote the development of the express delivery industry. The primary advantage of factor analysis is that it can solve the correlation problems between many variables, as far as possible preserving the results from indicators with special information. Factor analysis has practical significance for evaluating the development level of the express delivery industry, and has guiding implications for the industry's future development. Nevertheless, due to the orientation of the data selection in the research, it will unavoidably cause slight deviations in results, which should be improved in future explorations.

References

Cai, B., Yao, Z. Q., & Zhang, Q. (2008). The application of principal component analysis in the developmental status of economy of Jiangsu province. *Mathematics in Practice and Theory* (11), 7–16.

Feng, H. (2007). *The research on Chinese regional industrial competitiveness: Theory and empirical analysis* (Master dissertation, Jilin University).

Jia, J. K. (2009). Analysis of the relationship between principal component and factor analysis. *Knowledge Economy* (7), 168–169.

Li, L. (2012). *Research on national port price mechanism based on DEA model* [Master dissertation, Shanghai Jiao Tong University].

Liao, H. M., Liao, X. H., & Chen, H. K. (2015). Study on highway flood fuzzy probability comprehensive evaluation model based on method of factor analysis. *China Safety Science Journal, 25*(2), 141–146.

Sun, L. R. (2005). The principle and operation of structural equation modeling (SEM). *Journal of Ningbo University (Educational Science Edition)* (2), 31–34+43.

Wang, Y., & Cheng, F. X. (2012). Research and effect evaluation of resource marketing strategy model based on structural equation. *Contemporary Economics* (12), 134–135.

Wei, J., & Wang, J. (2011). Research on urban logistics capacity evaluation based on DEA and BP neural network. *China Market, 28*, 21–22+25.

Wen, X. R., Jiang, X. L., Zheng, Y., Xie, X. J., Li, C. G., & Yu, G. H. (2014). Evaluation of stand volume and related factors of *Pinus massoniana* artificial forests. *Journal of Central South University of Forestry & Technology, 34*(11), 32–36.

Yang, Y., Jin, Y. M., Sun, X. Y., & Chen, J. (2005). The source of atmospheric particulate matter in Fushun was analyzed by factor analysis. *Environmental Protection and Circular Economy* (4), 7–9.

Chapter 7
Operational Mechanism of Express Delivery Networks

This chapter compares the development, current operational status, and operational requirements of express delivery networks, then constructs a theoretical framework for the operational mechanisms of express delivery networks and deeply investigates their operational mechanisms.

7.1 Definition and Composition of Express Delivery Networks Operational Mechanism

The concept of operational mechanisms was initially applied to enterprise management, and then was subsequently introduced into many other areas in order to analyze problems. Through combining internal and external factors, the operational mechanism keeps the entire system in regular operation and circulation. It refers to the structural relationships and operating modes between various elements, in which the various components of the system are interlinked (Xu 2003).

The definition of the operational mechanism of express delivery network (hereinafter referred to as operational mechanism) in this book is: the structural composition and mutual relationship between the functional elements constituting the express delivery network, as well as the action process and operational form of each element, which mainly reflect the dynamic relationship of the interaction and mutual constraints between the components in the operation of the system. According to the concept of express delivery network, the operational mechanism is divided into three layers: physical, business and control. The physical layer is mainly composed of a motivation mechanism, which is the fundamental element of the operational mechanism; the business layer includes conduction, distribution and organizational management mechanisms, which play a normative role in the operational mechanism; the control layer mainly includes facilitation and guarantee mechanisms, playing the

P. Li and J. Wu, *Research on Innovation of Express Delivery Network Management*,
https://doi.org/10.1007/978-981-99-8982-9_7

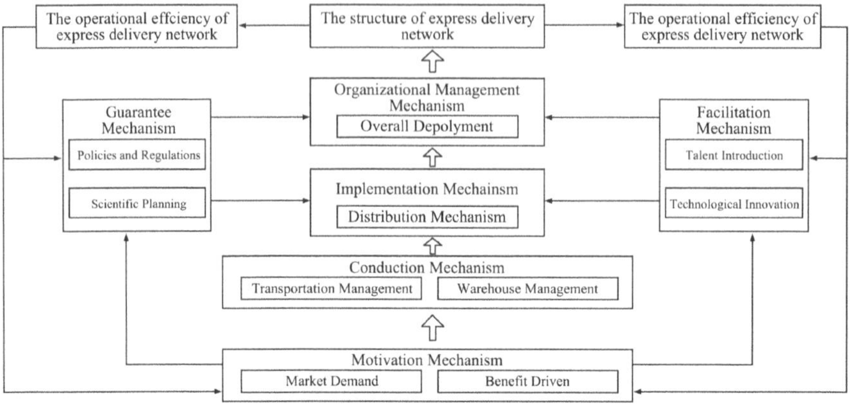

Fig. 7.1 The theoretical framework of the express delivery network operational mechanism (*Source* The Author)

role of supervision, control and coordination in the operational mechanism to ensure the smooth functioning of the network.

The operational mechanism includes one main route and two auxiliary routes, which correspond to an inner cycle and two outer cycles, respectively, with each cycle developing in conjunction with the others. The main line of the operational mechanism is composed of five elements: a motivation mechanism, a conduction mechanism, a distribution mechanism, an organizational management mechanism, and a structure of an express delivery network, which constitutes the inner cycle of the express delivery network operation. One side line is composed of three elements: the structure of the express delivery network, the operational efficiency of the express delivery network, and a facilitation mechanism, which link with the inner cycle to form an outer cycle. This outer cycle improves the operational efficiency of the express delivery network by optimizing and adjusting the structure of the express delivery network and by improving science and technology. Another side line is composed of three elements: the structure of the express delivery network, the operational efficiency of the express delivery network, and a guarantee mechanism, which form another outer cycle with the inner cycle. This cycle stably and orderly guarantees the operation of the express delivery network through policy regulation and legal constraints, as well as other guarantee mechanisms. The theoretical framework of the express delivery network operational mechanism is shown in Fig. 7.1.

7.2 Motivation Mechanism

In order for the network to operate effectively, a more sustainable and long-term motivation for nodes is needed, incorporating both an interest mechanism and a cost mechanism.

7.2.1 Interest Mechanism

The development of the express delivery network is derived from the persistent increase of market demand, which is the motivating force behind the swift growth of express delivery companies and the creation of interests. Interest is the fundamental motivating force and final justification for this. The express delivery network comprises of different subjects, and due to the discrepancies in the interests and objectives sought, there are both complementary interests and conflicts among them, and the cooperative or competitive relationship between them is continually fluctuating. Consequently, the way to coordinate the circulation of interests among the numerous subjects and devise a relatively stable interest circulation system to achieve an equilibrium between the collective interests and individual interests of the industry, is the result of the wagering among the subjects of the express delivery network system in the cooperation and competition.

The indicators of the business volume of the express delivery industry from 2010 to 2017 are presented in Fig. 7.2. In 2016, China's gross business volume exceeded 30 billion pieces, reaching 31.28 billion pieces with a year-on-year growth of 51.33%. This growth rate was seven times that of the GDP in the same period, thus indicating that the express delivery industry is at the forefront of modern service industries. In 2017, China's express delivery industry's business volume achieved a new peak, surpassing 40 billion pieces, with a year-on-year growth of 28.07% to reach 40.06 billion pieces.

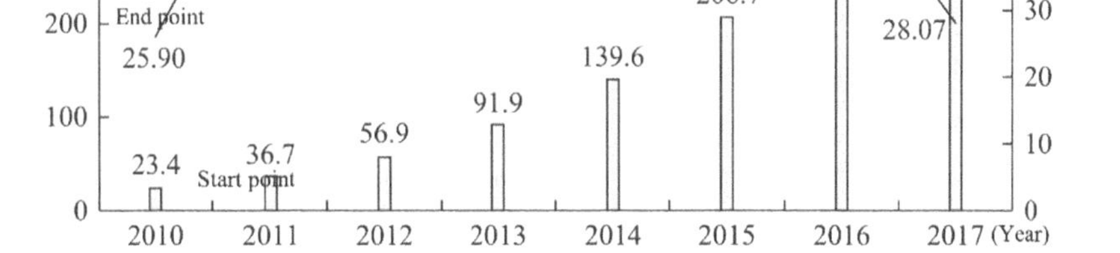

Fig. 7.2 The indicators of the business volume of the express delivery industry from 2010 to 2017 (*Source* The Author)

7.2.2 Cost Mechanism

Professor Osamu Nishizawa of Japan first proposed the Iceberg Cost Theory of Logistics, which suggests that by using mechanisms and innovating hardware, logistics costs can be reduced. This introduced mechanism can improve the value of the whole circular production process. As illustrated in Fig. 7.3, China's investment in fixed assets in the express delivery industry from 2011 to 2017 has been steadily increasing; however, the relevant regulation system remains inadequate, leading to high operating costs for express delivery networks. Studies in western countries have shown that direct labor costs account for only 10% of total costs in enterprises, whereas 35% of expenses are associated with express delivery (Wu 2005). It is evident that cost reduction in the express delivery industry has a substantial, positive multiplier effect on profitability.

Between 2010 and 2017, China's express delivery industry saw a steady increase in business volume and revenue. This sustained growth is indicative of a sharp rise in demand for express delivery services. Moreover, with the rapid proliferation of e-commerce and the emergence of artificial intelligence, enterprises in the express delivery industry have experienced considerable growth in recent years, contributing to a larger proportion of the country's Gross National Product (GNP).

7.3 Conduction Mechanism

The motivation mechanism provides a driving force for the express delivery network operation. However, it can only effectively and efficiently exploit this driving force when it is combined with the conduction mechanism. The conduction mechanism

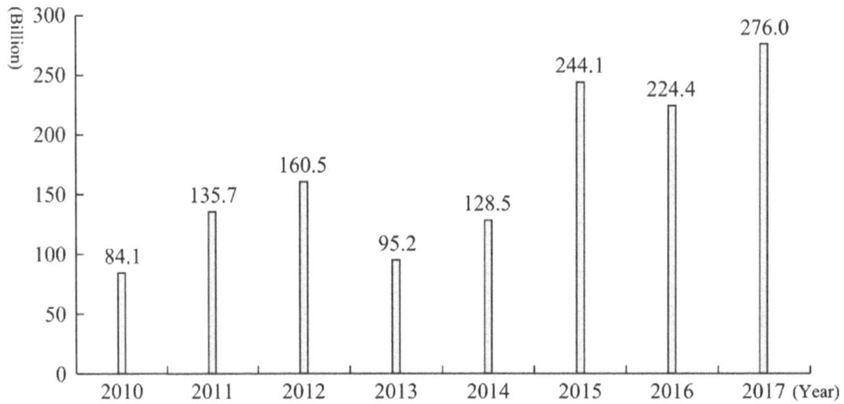

Fig. 7.3 China's investment in the fixed assets in the express delivery industry from 2011 to 2017 (*Source* The Author)

acts as the vehicle and tool for organizations to achieve network operation. Generally, shipping, storage, sorting, loading and unloading are the conduction components of express delivery network operation.

7.3.1 Transportation Management Mechanism

Shipping is the artery of modern logistics. Before modern logistics emerged, people associated shipping with logistics, for the reason that transportation was responsible for the majority of logistics tasks and created the time and space efficiency of packages. Currently, there are five primary shipping modes: railway transportation, road transportation, waterway transportation, air transportation and pipeline transportation. Each mode has its own unique advantages, disadvantages and applicable scope, as seen in Table 7.1.

Reasonable modes, tools and advanced technology in transportation can not only improve efficiency, but also avoid capacity waste that increases shipping time and

Table 7.1 Characteristics of various transportation

Mode of transportation	Advantages	Disadvantages	Scope of suitable application
Railway	It has large volume, low transportation cost, and strong continuity	It has high investment, long construction period	It is suitable for transportation of large-volume, long-distance and low-value items
Highway	It is flexible and available for door-to-door service	Its carry capacity is low	It is suitable for transportation of medium and short distance, medium and small quantity, high-value goods
Waterway	It has low freight with high volume	It is slow and susceptible to natural conditions	It is suitable for transportation of long distance with low value and high volume goods
Aviation	Its speed is fast and It is flexible. The requirements for cargo packaging is low	It has low transportation volume and high cost,and also be limited by weather	It is suitable for transportation with valuable and time-sensitive goods
Pipeline	It has large volume, low freight and stability in transportation	It has poor flexibility and high specificity	It is suitable for transportation of one-way, fixed-point and large-volume liquid cargo

Source The Author

overruns caused by unreasonable shipping such as transportation with no packages, roundabout, repeated, and improper selection. By adopting reasonable modes, routes and distribution schemes, the express delivery network can improve the actual load rate and transportation efficiency; thus, goods can be delivered to the destination in a timely and accurate manner, which enables the express delivery network to reduce transportation costs, save energy and improve operational efficiency.

7.3.2 Warehouse Management Mechanism

With the development of China's economy, the value of the express delivery network system has become increasingly apparent. The coordination of the express operational system is essential for the express delivery industry to facilitate the rapid development of China's economy and society. However, from the perspective of the current situation, the cost of the express delivery industry in China is high, and the cost reduction is directly related to the improvement of operational efficiency of the express delivery networks. In the current system, warehousing management is essential, and is a major cost factor. Therefore, it is necessary to formulate a suitable warehousing management mechanism. The general process of warehouse management is illustrated in Fig. 7.4.

Warehousing does not alter the function, nature and value of the items stored, but rather increases their value by preserving and perpetuating them. It adopts effective storage positioning technology, modern storage maintenance technology, and an

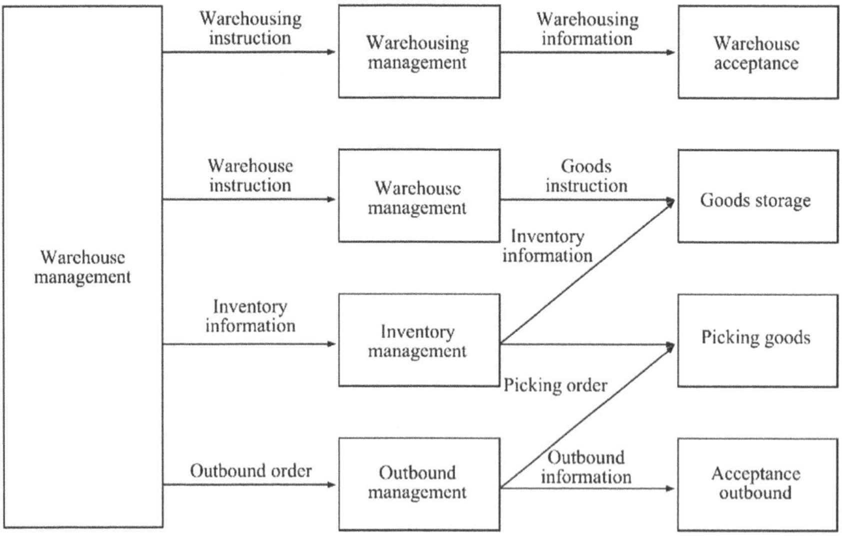

Fig. 7.4 The general process of warehouse management (*Source* The Author)

orderly warehouse management system to reduce time spent searching and accessing goods, accurately monitor the actual situation of goods storage, and facilitate the circulation of goods and capital to avoid unnecessary storage. Scientific warehousing not only guarantees the quality of the goods, but also increases inventory turnover and the utilization rate of warehouse facilities, reduces energy consumption during storage, and significantly improves the operational efficiency of the express delivery network.

Reasonable and effective warehouse management methods can be categorized into three main types.

7.3.2.1 Updating Facilities and Equipment

Express delivery enterprises should replace their outdated handling and sorting equipment, such as stackers, in order to adapt to the modernized trend of warehousing and improve the automation and overall management efficacy of the warehouse.

7.3.2.2 Improving Technical Proficiency

Express delivery enterprises should strengthen their introduction and research of warehousing technology, and utilize advanced science and technology such as information technology, telecommunication technology, and GPS technology to modernize the warehousing industry. This is to facilitate the transformation from manual operation management mode to electronic, digital, networked, mechanized, and automated operating mode.

7.3.2.3 Establishing Information Management System of Warehousing

Judging from the current situation, China has not yet implemented a unified warehouse management system. It is essential to note, however, that enterprises must create an informationalized platform to enhance the efficiency of warehousing utilization. Establishing and improving a networked information platform for warehouse management on a national level is necessary in order to achieve information-based warehouse management in China, allowing for smoother communication between enterprises. Through the timely sharing of relevant information, enterprises will be able to reduce their logistics costs and improve their economic benefits.

7.3.3 Packaging Management Mechanism

Packaging, designed to protect products and facilitate storage and transportation, can not only prevent damage incurred during loading, unloading, shipping and storage, but also damage caused by chemical changes and harmful organisms, thus effectively protecting products. As a low-profit industry, the express delivery industry needs to reduce costs and increase profits through focusing on cost control for each order. Currently, the commonly used packaging materials for express delivery services are packaging bags, cartons, and various fillers, which waste resources and pollute the environment. Through promoting a culture of saving and innovating in express packaging technology, we can reduce costs and save considerable social resources.

7.3.3.1 Packaging Laboratory

The Packaging Laboratory has two areas of research and development (R&D): Firstly, reducing the material costs by shrinking the size and area of packaging, thus decreasing wastage of resources; secondly, striving for energy savings and environmental protection by utilizing more eco-friendly materials for the recycling of packages.

7.3.3.2 Technology of Intelligent Packaging Algorithm

In November 2016, Cainiao's Network algorithm experts utilized big data and large-scale optimization technology to launch the "Intelligent Packaging Algorithm". This technology is capable of automatically computing the size, number, and packaging method of boxes necessary for goods when an order is generated, thereby optimizing the utilization of packing boxes with the aid of an algorithm. Furthermore, with proficient packaging, it is estimated that up to 0.5 billion kilograms of cartons can be saved per year.

Manual packaging traditionally relies solely on human experience, making it difficult to accurately estimate the volume and weight of the goods. In contrast, an "intelligent packaging algorithm" can quickly and accurately calculate the properties, quantity, weight, volume, and even the packaging methods of the goods when the order is generated, and match the box attributes to determine the suitable size and quantity of box. Compared to manual packaging, this can save more than 5% of packaging materials.

Currently, this technology is being tested and implemented in Cainiao's Station warehouse. According to the practical outcomes, the "Intelligent Packaging Algorithm" is significantly more cost and time-efficient than manual packaging.

7.4 Distribution Mechanism

The ultimate objective of the express delivery network is to fulfill the final stage of distribution. The express delivery network utilizes various distribution methods according to its own conditions and features to attain this supreme aim.

7.4.1 Self-Operated Distribution Mechanism

Self-operated distribution is a process in which express delivery networks organize and manage their distribution activities to meet commodity distribution needs. As the express delivery network is the subject of this distribution, it can access distribution resources more easily, facilitate internal communication, and improve distribution quality and timely response. This is beneficial for communication between distribution enterprises, as it enables them to create multiple large-scale logistics distribution centers to expand distribution. However, the "small and complete" distribution mechanism can cause an unreasonable use of social resources when the distribution scale is small. Establishing a distribution system on its own requires a large investment and has a long pay-off period, which makes the enterprises' centralized construction of core competitiveness insufficient, leading to higher distribution costs (Ren 2014).

China's large-scale manufacturing and distribution enterprises, predominantly franchised enterprises, typically employ a self-operated distribution model. Generally, larger group enterprises are advised to utilize self-operated distribution models, such as those employed by chain enterprises.

7.4.2 Distribution Mechanism of Third-Party

Third-party distribution, also referred to as outsourcing distribution, is typically associated with "first-party" seller logistics and "second-party" buyer logistics. The third-party distribution operating mechanism involves a third-party undertaking part or all of the distribution industry in order to complete all or part of the logistics activities of the enterprise. This type of distribution is characterized by its higher degree of logistics specialization, and is seen as a key way to effectively realize logistics socialization and rationalization. The general flow chart of the third-party distribution mechanism is presented in Fig. 7.5.

The continuous development of e-commerce and the third-party distribution system has led to this mechanism becoming the preferred distribution mechanism for e-commerce websites and industrial and commercial enterprises. The express delivery network can access high-grade logistics services of high quality, professionalism, and convenience, provided by the third-party logistics distribution mechanism, which decreases the cost of constructing logistics and increases the core competitiveness of the enterprise. The third-party distribution mechanism is of

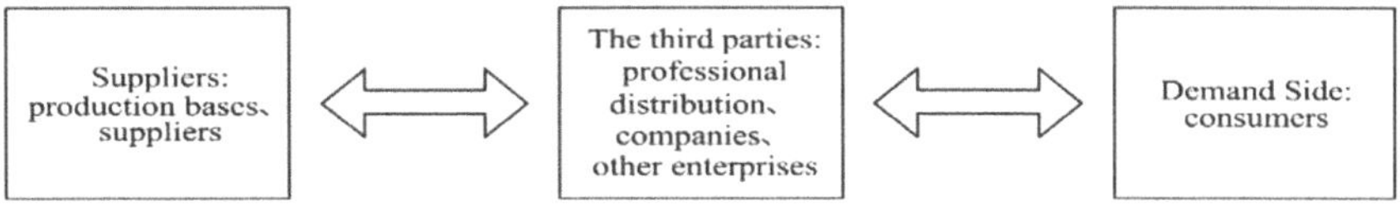

Fig. 7.5 The general flow chart of the third-party distribution (*Source* The Author)

immense importance in satisfying consumer demand, minimizing shortage costs, reducing delivery cycles, boosting consumer satisfaction, forming corporate image, and improving corporate reputation. Integrating logistics through the scale advantage of third-party logistics providers can help to control production operations and strike a balance between production and logistics, ultimately reducing logistics costs and the overall cost of the express delivery network. It can also actively integrate surrounding resources to maximize resource utilization rate and conserve social capacity. However, there are some inherent drawbacks to the third-party distribution mechanism. By outsourcing part or all of its distribution business, the express delivery network is unable to directly control customers in the same way as if its operations were self-managed, which could lead to customer loss (Li 2014).

7.4.3 *Interoperable Distribution Mechanism*

The interoperable distribution mechanism is a mechanism in which express delivery networks enter into contractual agreements in order to access the other party's logistical resources for their own benefit. Its main advantage is that it allows the express delivery network to expand its service scope and scale without additional capital, manpower, or equipment investments. However, the stability of this distribution mechanism could be improved, and efficient communication between networks is required in order to achieve a high level of organizational coordination and management. This mechanism has high requirements for logistics information technology, and is more suitable for transaction methods in modern, convenient logistics information conditions. An example of this is shown in Fig. 7.6.

The interoperable distribution mechanism seeks to enhance the distribution capabilities of the express delivery network, with its own services being the focal point; there are no specific requirements for cooperators, which can assume the form of

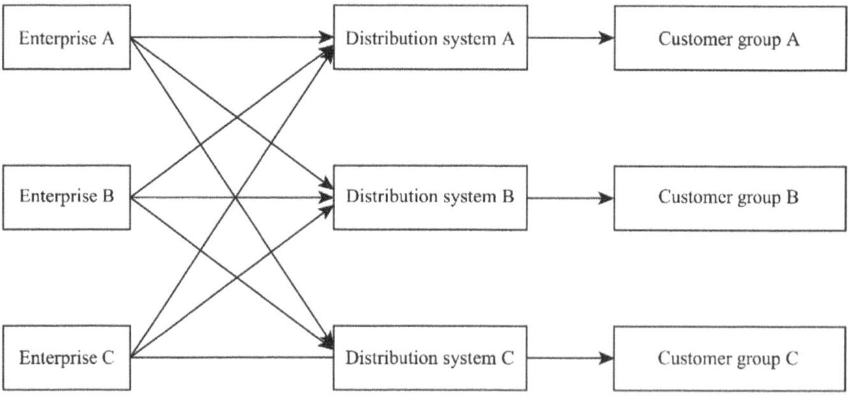

Fig. 7.6 The basic form of the interoperable distribution mechanism (*Source* The Author)

either distribution business enterprises or non-operating business enterprises. As the interoperable distribution mechanism itself is of primary importance, it is essential that the cooperation between parties is stable.

7.4.4 Collaborative Distribution Mechanism

The Collaborative Distribution Mechanism is an alliance established by an express distribution network, in which the unified planning and dispatching of the distribution center is used to improve distribution efficiency, realize distribution rationalization and optimize logistics resources. There are two main operating models for this mechanism: (1) the express delivery network reasonably arranges delivery times and routes according to the different requirements of each customer, and makes reasonable deliveries to each user; (2) in the delivery link, goods to be transported by various enterprises are loaded onto the same vehicle and then delivered to the designated location according to the specific requirements of the user. The general flow chart of the distribution mechanism is shown in Fig. 7.7.

The Collaborative Distribution Mechanism has a number of advantages. For the transporter, it enables the requirements of various customers in a given area to be managed in a unified way, as well as facilitating centralized and large-scale delivery. In addition, it reduces the unnecessary competition between various express delivery networks, allowing them to lower costs, enhance their distribution capacity and scale, and focus on core business. From a societal perspective, this mechanism effectively increases the actual carrying capacity of vehicles, reduces the total number of distribution vehicles, improves traffic conditions, and reduces total carbon emissions.

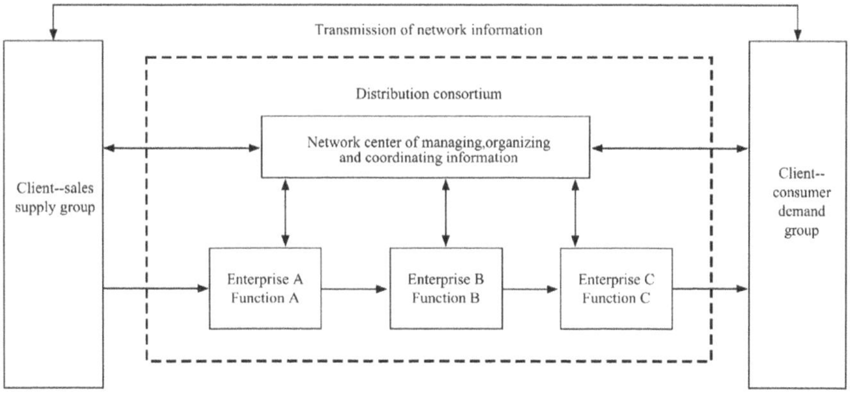

Fig. 7.7 The General flow of Collaborative distribution (*Source* The Author)

7.5 Organization and Management Mechanism

The efficacy of the organizational framework for the development of express delivery networks is dependent upon an effective organization and management mechanism. According to institutional economics, capital, human resources, and natural resources are not the primary catalysts of economic growth. Rather, it is the organizational system that determines the operational and developmental status of an organization, and serves as a productive force. Thus, the construction and design of organizational institutions is deemed crucial. The management mechanism of express delivery networks can be classified into three categories: self-operated mechanism, third-party mechanism, and fourth-party logistics mechanism.

7.5.1 *Self-Operated Logistics Mechanism*

The self-operation of the express delivery industry is primarily the integration of production and operations. In self-operated logistics, express delivery networks are equipped with warehouses, vehicles, and other logistic settings, as well as a management department to oversee the enterprise. The benefit of self-operated logistics is that they have control over their rights. The enterprise can solve problems in logistics management at a faster rate, and the information exchange between suppliers and customers is also up-to-date. By obtaining information, businesses can adjust their strategies in a timely manner. When it comes to cost budgeting, companies may not be able to acquire accurate supplier data when opting for self-operated logistics for delivery. In the internal control of raw materials and product sales, companies do not need to be worried about commission ratios in warehousing and after-sales service, thus eliminating transaction uncertainty and reducing risks. Furthermore, self-operated logistics of enterprises can also enhance their brand value and influence.

The self-construction of the logistics system can benefit customers by allowing them to obtain client's information and enables express delivery enterprises to adjust their strategies to meet customer needs in a timely manner. However, there are several drawbacks to running a self-operated business, including a low degree of specialization, large investments, and numerous management challenges.

7.5.2 *Third-Party Logistics Mechanism*

Third-party logistics (3PL) generally refers to the use of external resources to provide production services and the outsourcing of certain activities to third parties by enterprises. It is aimed at the first- and second-party, engaging in specialized logistics services through cooperation between them. It does not own the product or take part in the sale of goods. 3PL can help companies focus on their core business.

7.5.3 Fourth Party Logistics Mechanism

The composition of fourth-party logistics (4PL) includes entrusted customers, third-party service providers, management consulting firms, and value-added service providers, among others. 4PL can help enterprises reduce management and operational costs by enabling them to focus on their core business and outsource non-core aspects of logistics. This, in turn, enhances the impact of their core business and reduces associated costs, while simultaneously reducing capital occupation and effectively controlling storage costs. 4PLs have obvious professional advantages, providing customers with numerous resources that can deliver professional services and reduce operational expenses. Additionally, 4PLs benefit from network advantages, being reliant on networks and information, and large-scale network coverage can help customers extend their reach into overseas markets, thereby cutting operational logistics costs.

Due to their differences in many characteristics from different transmission mechanisms, there are variations in the management efficiency of express delivery networks, the efficiency of express delivery, the efficiency of information flow, and the efficiency of resource conversion. When geographical dimension is incorporated into different logistics mechanisms, the complexity of industrial operations is greatly increased in addition to internal operational problems, regional economic development level, population distribution, geographical characteristics, development level of related industries, people's consumption concept, etc. All these factors affect the operational efficiency and income. Operational efficiency to a certain extent determines the industry's development and income. High operational efficiency and high resource utilization will reduce costs and increase output; therefore, enterprises will consider further expanding regional investment to obtain more lucrative rewards, thus activating the vitality of the local express delivery industry. If the efficiency is low, in order to save costs and reduce losses, express delivery network will reduce the investment of resources, and if it continues to lose for a long time, it may even cancel the regional business.

In summary, the express delivery industry is a logistics distribution system which is based on an express delivery network and consists of components such as infrastructure, personnel, fixed asset investment, logistics distribution tools, transit stations and distribution points, related hardware facilities, technical means, and other elements. All of these components work together to ensure the normal operation of the express delivery industry. Moreover, these components are interdependent and interact with each other, which means that the allocation of elements and their arrangement will create a "butterfly effect" on the overall efficiency of the industry. Additionally, external factors such as the development of the macro economy and the advancement of science and technology can have a positive impact on the productivity of the express delivery industry. However, to maximize efficiency, more than just cutting-edge technologies and advanced management methods are needed; coordination between the various elements is also essential.

7.6 Facilitation Mechanism

In the operation of express delivery networks, facilitation mechanisms are critical in providing better support. Technological advances in express packaging, sorting, delivery robots, and unmanned aerial vehicles (UAVs) have proven to be highly effective in promoting the operation of express delivery networks.

7.6.1 Upgrading of Express Delivery Network Structure

Since China's reform and opening up, its economy has developed rapidly and its people's living standards have improved, resulting in an increased demand for safe and speedy express delivery services. The implementation of a comprehensive transportation network and the rapid development of computer technology have provided support for the express delivery industry, aiding the quick transmission of goods and information, enhancing logistics, information flow and capital flow efficiency, and decreasing the operating costs of the social economy in turn, thereby contributing to GDP growth and regional economic development. As a result, the express delivery industry has become a new service industry that promotes national economic development and industrial structure upgrading, while further driving the growth of related industries, particularly acting as a crucial link and a significant foundation for the promotion of e-commerce.

The structure of express delivery network in China needs to be adjusted and optimized with respect to market structure and service. Concerning market structure, China's express delivery industry is mainly concentrated in the low-end market, where competition is intense and disordered. Very few express delivery enterprises are able to contend with international enterprises in the high-end market. Through resource integration, express delivery enterprises in our country have become influential corporations, and thus entering the high-end market is an essential part of the market structure optimization of the express delivery industry. In terms of logistics service, the added value of services can be increased by improving technological innovation and informatization, cultivating high-quality personnel, optimizing the current network, transportation capacity and equipment, and improving the professional, modern and integrated services. To fundamentally eliminate the pattern of high investment and high consumption, it is essential to not excessively expand the scale of enterprises and ensure modernization and informatization.

7.6.2 Mechanisms for Talent Recruitment and Cultivation

Schultz posits that the comprehensive concept of capital can be divided into two components: human capital and physical capital. He suggests that, similar to physical capital, human capital has an increasing contribution to economic growth (Yang 2002). The sustainable development of the express delivery industry is contingent upon talent; this can be achieved by combining high-quality talent with various production factors. According to the actual situation of their own development, express delivery enterprises should improve their talent introduction mechanism, talent incentive system and talent training system by integrating professional logistics technology and cultivating excellent logistics management talents, so as to facilitate the high-quality development of talent and provide power for the industry's development.

7.6.3 Mechanism of Technology Transfer

Technological advancements can facilitate process optimization and improved logistics equipment of express delivery networks. Logistics equipment is the material basis of express delivery network systems, and since 1970, it has seen rapid development in China. Not only has the total amount of various logistics equipment increased quickly, but the level of automation and informatization has also been improved to some extent. Logistics equipment has been used in all areas of logistics. However, in general, the development of logistics equipment in our country does not meet the requirements of the new logistics tasks of the new century. Improvements to the management of logistics equipment, as well as the need for high-end logistics equipment, has led to lower operating efficiency of express delivery networks. Stimulating scientific and technological progress can enhance the current logistics equipment, optimize operational flow, and improve the efficiency of both the working and operational processes of express delivery networks.

7.7 Guarantee Mechanism

The operation of express delivery networks is largely based on the market mechanism, however, it needs to be further empowered. Government support is essential in this regard and should provide regulatory mechanisms to ensure the stability of market institutions.

7.7.1 Information Network Mechanism

Courier information is a valuable resource that should be exploited to its fullest potential in courier activities. Therefore, it is necessary to construct an express delivery network information system by applying express information technology, thus forming a comprehensive information collection and processing, storage, and transmission system, and then effectively handling and utilizing various express information to achieve the modern express's objectives. Relevant express information technology means include automatic identification systems, electronic data exchanges, and intelligent transportation systems, among others.

7.7.2 Policies and Regulations Mechanisms

Policies and regulations provide a code of conduct that both individuals and corporate entities must adhere to; those who violate such regulations must be subject to legal repercussions. Consequently, the enforcement of laws and regulations is essential for the advancement of express delivery companies.

The Postal Law of the People's Republic of China (outlines the specific contents of postal services, while *The Several Opinions of the State Council on Promoting the Development of the Express Delivery Industry* provides a series of effective measures to facilitate the development of the express delivery industry. *The Regulations on the Administration of Prohibited Items* are also clear in terms of the requirements. *The Notice of the State Post Bureau on printing and distributing ' The Regulations on the Administration of Smart Express Box Delivery Services (Interim)'* has standardized many issues of public concern, such as the fact that packages marked as fragile or with obvious damage to the outer packaging cannot be delivered through the intelligent express box. However, the regulations do not make any specific reference to whether fresh food can be delivered to smart express boxes, and the division of responsibilities has yet to be clearly defined in the event of damage to express mail, whether caused by the express link or a problem after delivery. On December 30, 2011, the General Administration of Quality Supervision, Inspection and Quarantine of the People's Republic of China and the Standardization Administration of the People's Republic of China jointly issued a series of national standards for express delivery services," which were officially implemented on May 1, 2012. It stipulates that the courier service organization should compensate for the complete delays of shipments in accordance with the relevant regulations.

7.7.3 Scientific Planning Mechanism

On March 30, 2017, the State Post Bureau released the "13th Five-Year Plan" for the development of express delivery services in the Beijing-Tianjin-Hebei region, the Yangtze River Delta region and the Pearl River Delta region. This "13th Five-Year" express delivery plan also provided guidance for the coordinated development of regional express delivery services throughout the country.

The Thirteenth Five-Year Plan for the Development of Express Delivery Services in the Beijing-Tianjin-Hebei Region has clearly outlined the development goals of the Beijing-Tianjin-Hebei region for the next five years. Comprehensive plans should be made to complete the industrial transformation and construction of transportation facilities, improve social and people's livelihood, and create an integrated layout of Beijing-Tianjin-Hebei measures and complementary advantages. They adjusted structure of it in 2020 to maintain medium-to-high-speed economic growth, form a regional integrated transportation network, improve the quality of the ecological environment, enhance people's living standards, increase the income of urban and rural residents, and improve the equalization of basic public services.

The Thirteenth Five-Year Plan for Express Delivery Service Development in the Yangtze Delta Region emphasized that enterprises in the Yangtze River Delta should maximize their agglomeration advantages and expedite the installation of national-level reform and experimental innovation areas, high-efficiency demonstration areas and high-end service guidance areas. By 2020, the Yangtze River Delta's total express delivery service volume will have reached 26.5 billion pieces, accounting for 38% of the national express delivery service volume. The cumulative revenue of the express delivery industry has reached 331.3 billion yuan, accounting for 42% of the total express delivery revenue in the country. In the designated area, express delivery services between key cities and cities should be completed within 12 h and express tasks between non-key cities should be completed within a single day (24 h). Consequently, the delay rate and cargo damage rate of express shipments during transit are expected to be substantially reduced, along with the loss rate and complaint rate.

The Thirteenth Five-Year Plan for the Development of Express Delivery Service in the Pearl River Delta Region proposed to develop the Pearl River Delta region into a demonstration area for national express transformation and upgrading, a hub for international express delivery services, and a convergence point for world-renowned express delivery companies. By 2020, the goal is to establish outlets in every township and express delivery services in every village. Under the guidance of the State Post Bureau and the local party committees and governments, all efforts will be made to create a growth engine for the reform and development of the national express delivery industry and to strive to construct a new pattern of coordinated regional express development characterized by its uniqueness.

7.7.4 *Risk Defense Mechanism*

The development of China's express delivery industry has been increasing at an accelerated rate. While this promotes growth, it also brings greater uncertainties and more complex risks. Due to the low entry threshold, lack of quality staff, small businesses, and small development momentum, the express market continues to be in need of new services for online shopping. Therefore, the rise of online shopping has not only provided development opportunities for the express delivery industry, but also posed great challenges to its service capacity and carrying capacity. Institutional, technical, policy, natural, and market risks are all intertwined, making the development of the express delivery industry highly unpredictable. To address this, governments and enterprises have implemented a variety of risk management systems, measures, methods, and means to protect against risks in logistics and distribution, and reduce the severity of losses due to risks.

7.8 Summary

In this chapter, the express delivery network operation mechanism is defined, its connotations are clarified, and it is noted that this mechanism consists primarily of motivation, conduction, distribution, facilitation, organization, management, and guarantee mechanisms. The motivation mechanism consists of two parts, the profit-driven and cost mechanisms, which together promote the construction and operation of express operations networks. The conduction mechanism is comprised of three parts: transportation management, warehousing management, and packaging management, all of which contribute to the efficiency of express operations. The distribution mechanism consists of four parts: self-operated distribution, third-party distribution, fourth-party distribution, and joint distribution; this is the core link of express operations and is closest to the final customer, impacting user experience. The organization and management mechanism is an organizational guarantee for the development of the express delivery network. Facilitation mechanisms can provide better support for the operation of express delivery networks, while the guarantee mechanism serves as a constraint and specification of express delivery network operations.

References

Li, W. J. (2014). "Discussion on 7S Standardized Management Mode of Express Enterprise Transfer Center". State Post Bureau of The People's Republic of China; China Express Association; Conference on the Development of Chinese (International) Express Industry. State Post Bureau of The People's Republic of China; China Express Association, 8(100–107).

Ministry of Public Security of State Post Bureau, Ministry of State Security, "Joyfully Announces Regulations on the Administration of Prohibited Items" State Post Bureau of The People's Republic of China. Retrieved from http://www.spb.gov.cn/zc/flfgjzc_1/201612/t20161216_934646.html2017.08.24.

Postal Law of the People's Republic of China. State Post Bureau of The People's Republic of China [EB/OL]. Retrieved from http://www.spb.Gov.cn/zc/flfgjzc_1/201210/t20121026_159203.html.2017.08.24.

Ren, D. Y. (2014). Study on a new university on-campus express delivery mode under student autonomy, *Logistics Technology*, *33*(11), 71–73.

The State Council of the People's Republic of China. The State Council of People's Republic of China, "Services Opinions of the State Council on Promoting the Development of the Express Delivery Industry" [EB/OL]. Retrieved from http://www.gov.cn/zhengce/content/2015-10/26/content_10256.htm.2017.08.24.

Wu, J. H. (2005). Application of modern cost management methods in enterprise logistics cost management [Master dissertation, Southwestern University of Finance and Economics].

Xu, Y. D. (2003). Research on the operation mechanism of modern agricultural science and technology park [Master dissertation, Sichuan Agricultural University].

Yang, W. Y. (2002) On the Forming Method of Human Capital and the Construction of Its Operational Mechanism, *Modern Finance and Economics-Journal of Tianjin University of Finance and Economics*, *04*,45–50.

Chapter 8
Evaluation of Express Delivery Networks' Operational Efficiency Using Three-Stage DEA Model

In recent years, the express delivery industry has seen an expansion in scale; however, there is still room for improvement in terms of the user experience. Customers have expressed dissatisfaction with the efficiency of pick-up and the timeliness of packages. This industry is influenced by a variety of factors, both internal and external, including economic and environmental ones. But the operational efficiency of the express delivery network is the main limiting factor in its growth. At present, the development of the express delivery industry in various regions across China is highly uneven, with the majority of business concentrated in a few areas; the Yangtze River Delta, the Pearl River Delta, and the Bohai Bay comprise 40%, 29%, and 16%, respectively. In contrast, the express delivery industry in other regions of China is relatively small in terms of revenue and size. This chapter utilizes a three-stage Data Envelopment Analysis (DEA) model to analyze the operational efficiency of regional express delivery industry. The input–output model is used to compare and evaluate the utilization efficiency of express delivery industry resources in various regions, and to further analyze the reasons for the uneven development of express delivery industry in different regions in China.

8.1 Research Map

Based on the characteristics of express delivery networks, this chapter has selected appropriate indicators to establish an evaluation system for the operational efficiency of said networks. In the first stage, the selected indicators will be analyzed and explained, followed by the application of DEA to calculate and compare the comprehensive technical efficiency, pure technical efficiency, and scale efficiency of express delivery networks in different provinces of China. Subsequently, the efficiency values obtained were utilized to analyze the causes of gaps in efficiency between provinces in order to comprehend the functioning of the express delivery network. In the second

P. Li and J. Wu, *Research on Innovation of Express Delivery Network Management*,
https://doi.org/10.1007/978-981-99-8982-9_8

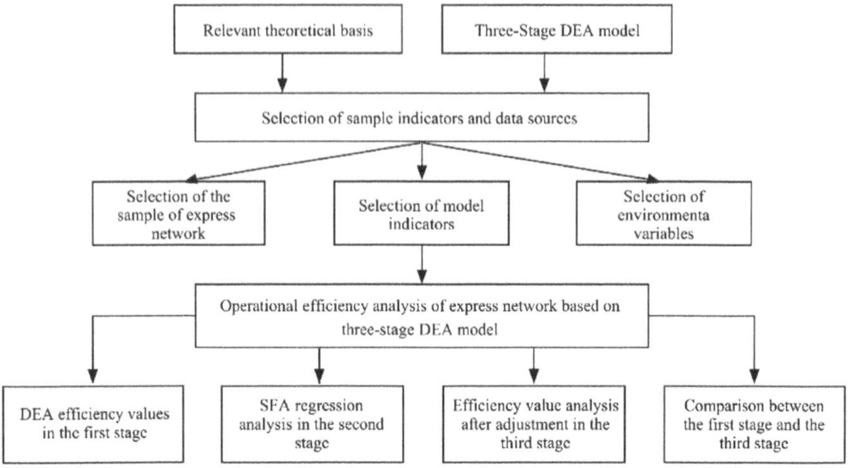

Fig. 8.1 Research map. *Source* The Author

stage, a Stochastic Frontier Analysis Model (SFA) algorithm was employed to reduce random errors and environmental factors. In the third stage, the final calculation results were achieved based on the revised input relaxation variable and compared with the calculated values obtained in the initial stage. According to the results of the comparative analysis, this chapter aims to evaluate the reasonableness of the relaxation variables adjusted by the SFA model in the second stage. In other words, it is assessed whether the Decision Making Unit (DMU) calculation results are more accurate and whether they can more precisely reflect the current operational efficiency of the express delivery network under the same environmental factors and random errors. This conclusion can provide guidance to provincial express delivery networks in China to address their operational shortcomings in accordance with the actual situation. The research map of this chapter is presented in Fig. 8.1.

8.2 Model Methods

8.2.1 DEA Model

In 1978, Charnes et al. proposed the DEA method as a non-parametric technique (Chen et al., 2010; Du & Huo, 2014) for evaluating efficiency in multi-input, multi-output Decision Making Units (DMUs). This model, referred to as CCR-DEA (Charnes et al., 1978), was then improved upon by subsequent scholars with the creation of the BCC model (Liu & Gao, 2010) and Super-efficiency model (Wang et al., 2009). The DEA model evaluates from the perspective of the most favorable

DMU by constructing a linear programming inequality group to determine the relative optimal weights of the input and output of DMU. This procedure does not require any assumptions to be made about the specific functional form of the DMUs' inputs and outputs, and the efficiency value of each DMU can be calculated according to the resulting weights. Of these DMUs, the relative efficiency value of the optimal one is the highest, at 1.

8.2.2 *Three-Stage DEA Model*

The Three-Stage DEA Model combines the traditional DEA Model and the Stochastic Frontier Analysis (SFA) Model, thereby avoiding the limitations of the simple first- and second-stage DEA Models. By building upon the conventional DEA Model, the Three-Stage DEA Model eliminates the negative effects of environmental factors and random errors on DMUs, clarifying the adjustment degree of DMUs, and enhancing the coordination between selected variables. The elimination of the aforementioned detrimental effects leads to the acquisition of more precise and valid efficiency values, thereby providing more accurate data for the efficiency evaluation of express delivery networks during operation.

The Three-Stage DEA Model, as its name suggests, consists of three stages. In the first stage, the traditional DEA models such as CCR and BCC are used to conduct efficiency analysis and to calculate the initial efficiency values of different DMUs as well as the input slack variables. In the second stage, the SFA model is employed to eliminate the negative impacts of environmental factors and random errors on the DMUs. At this point, appropriate environmental variables are selected according to discriminant requirements, and input slack variables are used for regression analysis. Finally, the third stage uses the modified input slack variables to evaluate DMU efficiency after the elimination of environmental variables and random errors has been completed. Figure 8.2 illustrates the steps of the Three-Stage DEA Model.

Suppose there are n DMUs, each denoted by DMU_j, $j = 1, 2, \ldots, n$. There are m types of inputs and s types of outputs, which are represented by $X_j = (X_{1j}, X_{2j}, \ldots, X_{nj})$ and $Y_j = (Y_{1j}, Y_{2j}, \ldots, Y_{nj})$ respectively, $j = 1, 2, \ldots, n.$. Then the dual programming $(D\varepsilon)$ of the Charnes-Cooper-Rhodes (CCR) model can be used to evaluate the effectiveness of the jth DMU. $(D\varepsilon)$ is:

$$\min\left[\theta - \varepsilon\left(\hat{e}^T s^- + e^T s^+\right)\right] = VD$$
$$\text{s.t.}\begin{cases} \sum\limits_{j=1}^{n} \lambda_j x_j + s^- = \theta x_0 \\ \sum\limits_{j=1}^{n} \lambda_j y_j - s^+ = y_0 \\ \lambda_j \geq 0,\ j = 1, 2, \ldots, n \\ s^- \geq 0,\ s^+ \geq 0 \\ \theta \in R \end{cases} \tag{8.1}$$

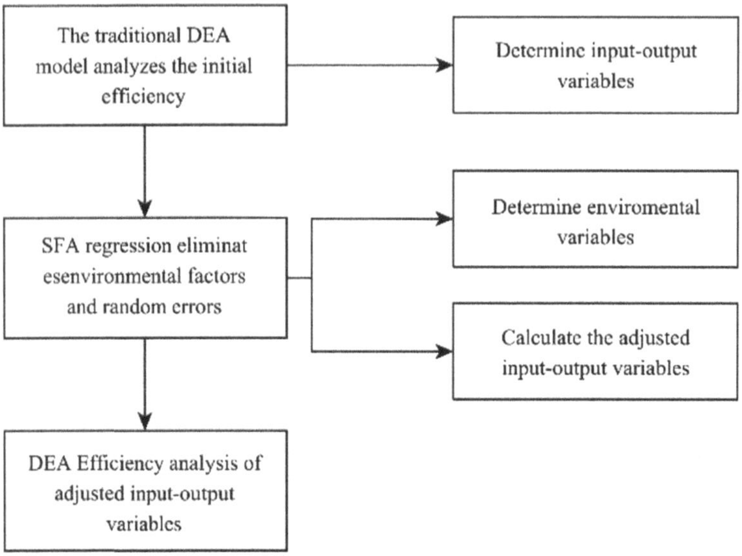

Fig. 8.2 Steps of three-stage DEA model. *Source* The Author

Among them, θ is the DEA efficiency value of DMU, and λ_i is the weight coefficient. s^+ and s^- represent the slack variable and surplus variable, respectively.

The CCR model mainly calculates the total efficiency value of Data Envelopment Analysis (DEA), but is unable to determine the values of pure technical efficiency and scale efficiency. The BCC model adds certain constraints to the CCR model in order to obtain the value of pure technical efficiency, thus increasing the effectiveness of the analysis. The scale efficiency value can be derived from the pure technical efficiency value θ_s calculated by the BCC model and the total efficiency value calculated by the CCR model.

If $\theta_s = 1$, the returns to scale of the DMU remain constant, a phenomenon known as effective scale efficiency.

If $\theta_s < 1$, then the scale of the DMU is invalid. At this point, it is necessary to determine whether the returns to scale of the DMU are increasing or diminishing. This judgment can be made by calculating the economies of scale $n\Sigma\lambda_i$ (Zhang & Che, 2011).

Assuming that the value of economies of scale is M, if $M = 1$, the economies of scale remain unchanged, indicating that the DMU is optimal. However, if $M < 1$, the DMU is subject to increasing economies of scale. The lower the M value, the greater the trend of increasing scale.

8.3 Establishing an Indicators System

8.3.1 Selection of Indicators

The selection of indicators will affect the accuracy of the empirical results when using a three-stage DEA model to calculate operational efficiency. Moreover, the number of indicators to be selected should also be taken into consideration. Despite this, existing research has rarely involved the selection principles of relevant evaluation indicators. Generally, decision makers choose evaluation indicators subjectively, leading to issues with the selection method and quality of indicators, which require further discussion. Thus, the following principles should be followed in the process of constructing the evaluation indicator system of the operational efficiency of the express delivery network:

1. The principle of accessibility. The selected indicators should be based on the premise that they are quantifiable, and that the data for them can be easily collected. The selected indicators should either be directly available in the database, or should be obtainable through corresponding calculations.
2. The principle of integrity. The establishment of a comprehensive indicator system is essential for a comprehensive evaluation of the target object. Consequently, this chapter need to design a system that can encompass all aspects of the goal, meeting both qualitative and quantitative requirements, and accurately depict reality.
3. Guiding principles. It is essential to establish an indicator system that can continuously optimize the express delivery network in order to ensure its efficient operation and sustainable development, with a role of guidance and monitoring.
4. Systematic principles. When selecting an indicator system, the principles of particularity and universality should be taken into account. Furthermore, the hierarchy and correlation of the indicators should be given particular importance when combining the key characteristics of research objectives.

The authors conducted a thorough investigation of existing research when selecting the indicator variables. Through comparative analysis, the authors established an input and output indicator system, which is shown in Table 8.1 and explained in further detail below.

Table 8.1 Selection of input, output and environmental indicators

Category	Indicators
Input indicators	Express delivery outlets, total length of post roads, the volume of express business
Output indicators	The income of express business
Environmental indicators	Average number of people served by each business outlet, infrastructure construction, the number of employees in the express delivery industry

Source The Author

8.3.1.1 Input Indicators

1. Express delivery outlets. Express delivery outlets are important fixed assets in the express delivery industry. In accordance with the service capabilities of the outlets, each outlet has a limited service area and a limited number of customers. Generally, the density of outlets is correlated with the regional demand for express delivery services. The more outlets there are, the higher the demand for express delivery services. The operational efficiency of the express delivery network is significantly impacted by the number of outlets. If the outlets are sparsely distributed, they cannot meet business needs and could lead to a decline in service quality, whereas outlets that are too densely distributed would lead to wasted resources and increased costs. Consequently, it is advisable to incorporate the number of express delivery outlets as an input measure when evaluating the operational efficiency of the express delivery industry.
2. Total length of postal routes. The total length of the postal route incorporates the lengths of the roads, railways, and postal air routes. This length reflects the construction of regional logistics infrastructure and the scale of the regional transportation industry. Express delivery networks are a unique functional form of transportation network. Compared to the length of transportation routes, the length of postal routes has a closer relationship with the express delivery industry, and is thus more suitable as an indicator for certain analyses of the express delivery industry.
3. The volume of express delivery business. The volume of express delivery business, an evaluation indicator in the logistics industry, serves as a reflection of express delivery service and regional economy, and is the most direct indicator of revenue for the express delivery business.

8.3.1.2 Output Indicators

Revenue from express delivery business is the primary output of express delivery industry operations, and is the most direct indicator of the operational status of the express delivery industry.

8.3.1.3 Environmental Indicators

The selection principle for environmental variables is that they have a significant impact on the operational efficiency of express delivery networks and the efficiency of DMUs beyond the range of subjective control. Environmental factors are stable and remain unaffected by the subjective will of DMUs. Given that many environmental factors influence the operational efficiency of express delivery networks, this chapter selects three environmental variables based on existing literature to analyze the impact of external environmental factors on the operational efficiency of express delivery networks.

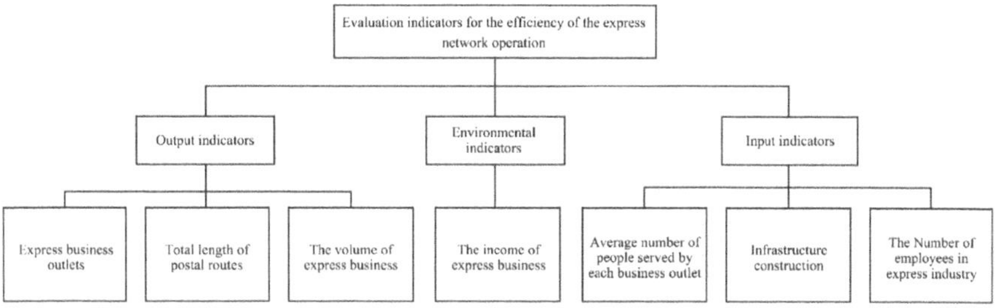

Fig. 8.3 An evaluation indicator system for the operational efficiency of the express delivery network. *Source* The Author

1. Average number of people served by each business outlet. Generally, the more efficient express delivery outlets have fewer customers.
2. Infrastructure construction. The economic environment of each enterprise varies. The more developed the economy of a region is, the more comprehensive the infrastructure construction is, and the more energy the express delivery industry can allocate to improving its efficiency, which is beneficial to the sustainable development of the express delivery industry.
3. The number of employees in the express delivery industry. Generally, the express delivery industry tends to develop more favorably in areas with a larger workforce, increased demand for skilled labor, and higher operational efficiency of express delivery services.

Based on the analysis of input and output indicators and environmental variables, and in accordance with the criteria of indicators such as accessibility, integrity, guidance, and systematicity, seven indicators have been selected for the construction of an evaluation indicator system for the operational efficiency of the express delivery network. This indicator system is divided into three categories, as depicted in Fig. 8.3.

8.3.2 Data Sources and Sample Selection

When employing the DEA model for analysis, the homogeneity of DMU should be assessed, reflecting whether the decision-making tasks of each DMU are consistent with the decision-making objectives. If the chosen DMU does not possess such homogeneity, it has no substantive value for the application of the DEA model.

When using the three-stage DEA model to evaluate efficiency, the number of indicators selected must meet the condition that the number of DMUs is more than twice the sum of input and output indicators; otherwise, the research results may be invalid. This study selected three input indicators and one output indicator for the express delivery industry in 31 provinces in mainland China from 2011 to 2017, thus satisfying the conditions for applying the three-stage DEA model. The data used are from the *China Statistical Yearbook* (2012–2018), as presented in Tables 8.2, 8.3, 8.4, 8.5, 8.6, 8.7, and 8.8.

8.4 Empirical Results and Analysis

8.4.1 Initial Efficiency with BCC Model

The operation of the express delivery industry can be viewed as a multi-input, multi-output decision-making model, and its resource utilization efficiency is affected by several factors, which can be measured from various perspectives. In the first stage of the model, the comprehensive technical efficiency, pure technical efficiency, scale efficiency and returns to scale of each province were empirically analyzed according to the input and output indicators of 31 provinces in China from 2011 to 2017, using the application of the BCC model in DEAP Version 2.1. The initial efficiency calculation results of the selected 31 provinces are presented in Tables 8.9, 8.10, 8.11, 8.12, 8.13, 8.14, and 8.15.

A line chart is used in Fig. 8.4 to observe the changes in comprehensive technical efficiency, pure technical efficiency, and scale efficiency.

The comprehensive technical efficiency of the express delivery network reflects its overall efficiency. From 2011 to 2017, the total efficiency values of the express delivery network were 0.691, 0.608, 0.625, 0.574, 0.616, 0.580, and 0.582, respectively. The average efficiency over the seven-year period was 0.611, indicating that the average production waste of the express delivery network during this time-frame was 38.9%. The overall efficiency of the express delivery network is not high, as indicated by the average pure technical efficiency and average scale efficiency being 0.670 and 0.924 respectively. Tables 8.9, 8.10, 8.11, 8.12, 8.13, 8.14 and 8.15 indicate that only two provinces in the sample, Shanghai and Xizang, have a total efficiency value of the operation network of 1. Consequently, the comprehensive technical efficiency of the express delivery network is not high and is low in several provinces, suggesting that the overall level of express delivery network operational efficiency is significantly differentiated.

As Fig. 8.4 illustrates, the fluctuation range of the comprehensive technical efficiency of express delivery network operations over seven years was small when the effects of environmental variables and random errors on the efficiency value were not taken into account. On the whole, the trend was stable, with a decreasing, increasing, and then decreasing pattern. The trend for both pure and comprehensive technical efficiency was similar, though the scale efficiency had only a minor alteration. The comprehensive technical efficiency was significantly impacted by pure technical efficiency.

Table 8.2 Input–output indicator data of express delivery industry in 31 provinces of China in 2011

Provinces	The income of express delivery business (ten thousand yuan)	Express delivery business outlets (number)	Total length of postal routes (km)	The volume of express delivery business (ten thousand pieces)
Beijing	625,532.3	2507	376,504	33,663.4
Tianjin	127,066.3	765	94,118	5130.6
Hebei	162,355.1	2461	71,367	8660.4
Shanxi	46,383.2	1300	106,655	2098.5
Neimenggu	45,339.8	1112	110,792	1995.0
Liaoning	151,316.8	2545	136,099	6211.5
Jilin	63,106.0	1429	91,456	2647.3
Heilongjiang	71,495.1	1654	114,962	3066.0
Shanghai	1,218,226.3	5208	259,717	40,914.9
Jiangsu	708,697.6	5837	211,468	38,509.0
Zhejiang	844,874.2	3965	260,054	49,660.8
Anhui	112,013.8	2260	65,219	6628.3
Fujian	316,623.4	2288	195,662	15,764.9
Jiangxi	66,607.0	2142	93,086	3716.1
Shandong	335,733.0	5506	302,217	18,439.0
Henan	140,529.4	4325	149,659	8377.9
Hubei	137,666.8	2569	191,578	8282.1
Hunan	114,983.0	2510	115,765	6342.7
Guangdong	1,632,100.4	7070	710,690	75,689.7
Guangxi	72,613.5	1941	131,334	3495.7
Hainan	20,164.6	632	93,432	953.5
Chongqing	76,828.8	2276	103,923	4068.3
Sichuan	177,908.0	3938	190,342	10,216.3
Guizhou	36,576.8	1637	99,155	1533.9
Yunnan	70,425.4	2134	178,708	3041.4
Xizang	11,750.0	225	25,098	284.3
Shaanxi	83,740.6	1842	237,536	3941.6
Gansu	83,740.6	1842	237,536	3941.6
Qinghai	8713.1	165	34,684	244.5
Ningxia	15,071.8	317	46,423	679.0
Xinjiang	57,706.2	1231	237,636	1919.6

Source The Author

Table 8.3 Input–output indicator data of express delivery industry in 31 provinces of China in 2012

Provinces	The income of express delivery business (ten thousand yuan)	Express delivery business outlets (number)	Total length of postal routes (km)	The volume of express delivery business (ten thousand pieces)
Beijing	762,737.0	3052	405,521	48,073.7
Tianjin	144,376.7	1068	157,480	6364.0
Hebei	214,740.6	2734	78,040	12,469.1
Shanxi	54,284.2	1728	97,314	2805.3
Neimenggu	58,718.6	1596	154,666	2440.0
Liaoning	179,310.6	3087	252,991	7757.4
Jilin	80,722.5	1601	179,250	3854.4
Heilongjiang	81,909.5	1881	138,855	3623.5
Shanghai	1,828,629.3	3792	192,241	59,905.3
Jiangsu	1,034,124.0	6405	240,630	63,870.5
Zhejiang	1,197,342.6	5155	449,588	81,986.8
Anhui	145,715.8	2434	139,364	9731.4
Fujian	420,985.5	3117	182,584	25,593.8
Jiangxi	84,144.3	2369	108,198	5472.6
Shandong	419,168.3	5173	309,689	24,731.8
Henan	192,209.6	4305	131,284	12,503.4
Hubei	183,033.8	4035	124,877	11,629.8
Hunan	164,752.9	2846	142,207	10,022.7
Guangdong	2,456,278.2	10,015	578,539	133,770.5
Guangxi	87,490.9	2304	273,528	4395.1
Hainan	23,230.1	675	94,378	1123.6
Chongqing	103,426.5	2506	110,951	5497.9
Sichuan	225,753.7	6217	269,222	12,814.4
Guizhou	40,386.8	1921	136,429	1801.0
Yunnan	85,357.5	2598	197,454	3774.4
Xizang	12,989.2	247	29,485	320.1
Shaanxi	101,893.6	2312	146,020	5085.0
Gansu	35,255.3	1697	126,312	1470.3
Qinghai	9844.1	241	72,785	286.7
Ningxia	52,709.0	364	54,058	2967.8
Xinjiang	71,803.8	1554	281,167	2406.0

Source The Author

Table 8.4 Input–output indicator data of express delivery industry in 31 provinces of China in 2013

Provinces	The income of express delivery business (ten thousand yuan)	Express delivery business outlets (number)	Total length of postal routes (km)	The volume of express delivery business (ten thousand pieces)
Beijing	937,978.8	3826	228,443	81,818.2
Tianjin	177,653.7	1405	448,060	8719.0
Hebei	288,768.9	3260	57,874	20,755.7
Shanxi	68,725.6	2727	73,422	8869.0
Neimenggu	64,417.0	2104	98,198	2839.0
Liaoning	227,256.9	3641	184,995	11,411.1
Jilin	87,529.9	2278	251,159	4526.7
Heilongjiang	100,134.0	2645	140,651	5393.9
Shanghai	2,575,921.3	4568	153,496	95,012.4
Jiangsu	1,429,899.2	8913	197,750	98,415.5
Zhejiang	1,797,804.2	6816	228,317	141,952.8
Anhui	195,051.8	2925	428,691	13,755.5
Fujian	615,612.7	4125	104,462	44,535.8
Jiangxi	129,791.0	3827	182,600	9751.5
Shandong	545,011.8	6153	81,476	31,375.8
Henan	265,020.9	5626	242,453	19,443.9
Hubei	285,487.2	5178	147,995	21,991.0
Hunan	195,896.5	4364	153,786	15,446.7
Guangdong	3,367,805.4	13,461	172,456	210,670.3
Guangxi	112,047.4	2764	594,990	6745.1
Hainan	28,075.7	864	291,560	2226.9
Chongqing	136,957.5	3044	90,861	10,614.8
Sichuan	304,253.8	8465	154,470	24,400.9
Guizhou	60,342.1	3070	270,825	2931.2
Yunnan	105,318.7	3095	129,375	6870.3
Xizang	14,847.2	303	275,486	378.8
Shaanxi	133,661.0	2529	52,302	9552.2
Gansu	40,780.6	2028	168,030	1788.5
Qinghai	11,903.5	725	183,447	417.5
Ningxia	24,309.7	541	63,879	972.9
Xinjiang	88,551.1	2381	45,723	5092.1

Source The Author

Table 8.5 Input–output indicator data of express delivery industry in 31 provinces of China in 2014

Provinces	The income of express delivery business (ten thousand yuan)	Express delivery business outlets (number)	Total length of postal routes (km)	The volume of express delivery business (ten thousand pieces)
Beijing	1,476,107.9	4110	425,591	111,011.9
Tianjin	250,712.6	1342	133,348	12,404.2
Hebei	410,725.8	3697	169,077	34,019.1
Shanxi	103,493.4	3568	55,370	9130.4
Neimenggu	103,218.8	2462	199,086	4363.6
Liaoning	300,127.3	3846	295,151	16,656.4
Jilin	130,649	2369	127,662	6640.2
Heilongjiang	124,136.3	2897	163,633	7014.5
Shanghai	3,613,060.4	4477	207,970	128,366.1
Jiangsu	2,010,744.2	11,861	383,630	148,435.2
Zhejiang	2,744,402.5	7217	501,588	245,744.8
Anhui	291,488.9	3786	134,963	23,859.1
Fujian	810,815	4222	192,848	65,417.3
Jiangxi	182,151.1	3360	85,252	15,993.6
Shandong	670,123.1	6498	322,284	44,685
Henan	408,429.9	6481	157,955	29,484
Hubei	413,805.5	5581	161,205	33,143.8
Hunan	262,122.8	4792	172,200	22,716.2
Guangdong	4,612,533.2	14,476	705,444	335,555.9
Guangxi	155,902	3370	234,766	9055.4
Hainan	43,190.6	1106	92,629	2248.6
Chongqing	201,060	3542	149,337	13,886.3
Sichuan	479,628	9266	251,631	37,941.8
Guizhou	98,170.6	3054	99,562	4669.1
Yunnan	152,401	3458	253,942	8546.1
Xizang	17,384.5	284	53,144	484.3
Shaanxi	179,548.3	3205	64,780	13,762.3
Gansu	51,198	2573	183,221	2655.6
Qinghai	15,442.2	425	83,865	579.8
Ningxia	33,500.2	642	12,208	1514.1
Xinjiang	107,313.3	2489	232,215	5940.5

Source The Author

Table 8.6 Input–output indicator data of express delivery industry in 31 provinces of China in 2015

Provinces	The income of express delivery business (ten thousand yuan)	Express delivery business outlets (number)	Total length of postal routes (km)	The volume of express delivery business (ten thousand pieces)
Beijing	1,816,522.5	6110	273,171	141,447.3
Tianjin	435,370.4	1814	110,688	25,624.4
Hebei	561,787.9	4830	170,096	54,911.9
Shanxi	152,901.4	5580	93,321	11,477.3
Neimenggu	123,053.4	3477	222,550	5410.1
Liaoning	395,949.5	5295	277,348	24,674.1
Jilin	169,634.0	4081	92,147	9017.0
Heilongjiang	213,466.8	4244	160,630	12,636.8
Shanghai	4,552,476.2	7781	242,349	170,778.0
Jiangsu	2,907,286.3	10,866	405,531	229,047.7
Zhejiang	3,838,082.6	11,829	618,999	383,145.9
Anhui	461,132.8	7039	150,960	39,935.6
Fujian	1,008,474.8	6453	203,778	88,786.2
Jiangxi	276,692.2	4284	96,390	23,471.8
Shandong	970,605.4	8599	369,063	73,424.9
Henan	631,077.0	7408	158,490	51,449.7
Hubei	595,633.7	9129	163,136	50,847.3
Hunan	338,934.3	7016	183,302	31,786.4
Guangdong	6,159,135.7	21,965	413,740	501,335.2
Guangxi	217,788.4	4834	251,304	12,540.9
Hainan	63,444.8	1513	96,833	2953.0
Chongqing	286,533.2	3685	191,486	20,525.4
Sichuan	628,897.8	11,833	255,526	48,796.6
Guizhou	132,428.3	4815	99,684	7034.3
Yunnan	201,152.5	4774	269,144	11,109.1
Xizang	16,826.3	888	56,211	578.2
Shaanxi	272,792.1	4335	208,078	20,351.0
Gansu	72,537.1	3835	192,349	3541.4
Qinghai	18,236.6	623	87,978	716.6
Ningxia	48,355.7	941	12,541	2231.9
Xinjiang	129,256.1	3194	259,607	7050.7

Source The Author

Table 8.7 Input–output indicator data of express delivery industry in 31 provinces of China in 2016

Provinces	The income of express delivery business (ten thousand yuan)	Express delivery business outlets (number)	Total length of postal routes (km)	The volume of express delivery business (ten thousand pieces)
Beijing	2,565,681.3	7156	597,079	196,029.0
Tianjin	634,879.9	1956	141,418	41,005.4
Hebei	942,582.7	6537	149,981	90,392.4
Shanxi	221,412.2	5972	96,086	18,665.2
Neimenggu	185,011.0	4150	266,580	8470.6
Liaoning	556,909.3	5937	278,423	39,825.9
Jilin	251,313.4	3748	126,118	13,894.0
Heilongjiang	331,634.7	4714	152,970	21,769.8
Shanghai	7,095,143.5	7629	234,650	260,274.4
Jiangsu	3,391,633.5	12,951	205,555	283,823.2
Zhejiang	5,412,544.6	11,948	694,120	598,770
Anhui	705,619.0	7528	61,376	68,878.3
Fujian	1,348,336.3	7418	227,354	128,985.8
Jiangxi	412,915.5	6561	94,041	38,304.6
Shandong	1,389,811.6	10,221	439,671	120,553.9
Henan	943,664.1	10,827	153,231	83,875.3
Hubei	871,650.5	11,812	178,815	77,348.1
Hunan	515,976.6	8004	96,802	48,603.5
Guangdong	8,802,789.8	21,602	448,236	767,241.6
Guangxi	338,879.1	6166	252,893	22,835.4
Hainan	100,338.6	1740	91,151	4869.4
Chongqing	389,617.2	4955	137,210	28,382.5
Sichuan	963,552.7	16,532	272,761	80,147.8
Guizhou	217,919.1	6074	100,813	11,260.1
Yunnan	289,569.0	6306	278,529	17,445.8
Xizang	20,713.0	965	57,598	734.4
Shaanxi	456,462.2	6435	209,355	36,901.6
Gansu	125,040.5	4635	192,557	6065.1
Qinghai	30,040.5	986	95,136	1078.6
Ningxia	58,591.2	1297	52,782	3241.5
Xinjiang	173,369.1	3946	201,760	8661.9

Source The Author

Table 8.8 Input–output indicator data of express delivery industry in 31 provinces of China in 2017

Provinces	The income of express delivery business (ten thousand yuan)	Express delivery business outlets (number)	Total length of postal routes (km)	The volume of express delivery business (ten thousand pieces)
Beijing	3,038,329.8	5016	565,619	227,452.1
Tianjin	763,314.9	2084	117,486	50,199.0
Hebei	1,264,880.7	7014	203,289	119,389.3
Shanxi	299,901.6	6596	89,303	24,359.1
Neimenggu	239,592.8	4605	282,129	11,035.5
Liaoning	680,585.6	7125	298,684	51,434.5
Jilin	304,484.0	4711	183,551	17,569.4
Heilongjiang	358,414.8	5818	177,532	23,185.6
Shanghai	8,688,851.6	16,374	189,648	311,503.7
Jiangsu	4,081,730.6	15,190	539,417	359,627.8
Zhejiang	6,682,204.0	26,487	758,228	793,231.1
Anhui	895,715.9	10,652	185,191	86,332.3
Fujian	1,619,683.4	9308	456,716	166,110.7
Jiangxi	491,976.7	10,238	90,254	43,754.5
Shandong	1,705,165.2	14,936	452,268	151,474.6
Henan	1,159,337.9	12,150	425,176	107,377.6
Hubei	1,190,450.7	14,222	199,008	101,277.9
Hunan	641,882.9	9810	170,706	59,181.6
Guangdong	11,466,893.4	24,234	2,009,200	1,013,468.0
Guangxi	448,672.2	7631	263,994	31,750.3
Hainan	126,970.8	1714	86,508	5915.8
Chongqing	447,311.3	6583	136,593	32,874.9
Sichuan	1,274,785.5	18,962	291,551	110,795.9
Guizhou	311,536.8	8047	148,484	15,781.9
Yunnan	360,114.9	7970	290,966	22,775.8
Xizang	20,487.7	1073	64,694	567.5
Shaanxi	563,581.1	8466	169,302	45,750.7
Gansu	148,095.9	4863	188,054	7201.7
Qinghai	38,832.9	970	86,279	1449.7
Ningxia	67,784.4	1319	56,489	3721.5
Xinjiang	189,518.9	3767	208,349	9042.4

Source The Author

Table 8.9 Full efficiency productivity index of express delivery network resources in 31 provinces of China in 2011

Provinces	Comprehensive technical efficiency	Pure technical efficiency	Scale efficiency	Returns to scale
Beijing	1.000	1.000	1.000	–
Tianjin	0.821	0.857	0.959	irs
Hebei	0.624	0.762	0.819	irs
Shanxi	0.613	0.691	0.887	drs
Neimenggu	0.619	0.710	0.872	drs
Liaoning	0.762	0.802	0.950	drs
Jilin	0.706	0.761	0.928	drs
Heilongjiang	0.682	0.749	0.911	drs
Shanghai	1.000	1.000	1.000	–
Jiangsu	0.714	0.760	0.941	irs
Zhejiang	0.902	0.909	0.992	irs
Anhui	0.558	0.648	0.817	irs
Fujian	0.668	0.669	0.997	drs
Jiangxi	0.553	0.574	0.964	drs
Shandong	0.584	0.607	0.961	drs
Henan	0.534	0.552	0.968	drs
Hubei	0.517	0.546	0.947	drs
Hunan	0.576	0.593	0.972	drs
Guangdong	0.955	1.000	0.955	drs
Guangxi	0.608	0.668	0.910	drs
Hainan	0.541	0.595	0.909	drs
Chongqing	0.581	0.609	0.955	drs
Sichuan	0.553	0.576	0.960	drs
Guizhou	0.628	0.730	0.859	drs
Yunnan	0.623	0.743	0.838	drs
Xizang	1.000	1.000	1.000	–
Shaanxi	0.595	0.687	0.866	drs
Gansu	0.591	0.725	0.816	drs
Qinghai	0.907	1.000	0.907	irs
Ningxia	0.622	0.673	0.925	irs
Xinjiang	0.776	0.954	0.814	drs
Median	0.691	0.748	0.923	–

Note "irs" stands for rising, "drs" stands for falling, and "–" stands for unchanged. The same below
Source The Author

Table 8.10 Full efficiency productivity index of express delivery network resources in 31 provinces of China in 2012

Provinces	Comprehensive technical efficiency	Pure technical efficiency	Scale efficiency	Returns to scale
Beijing	0.520	0.561	0.927	irs
Tianjin	0.709	0.728	0.974	drs
Hebei	0.558	0.610	0.915	irs
Shanxi	0.568	0.597	0.951	drs
Neimenggu	0.645	0.746	0.864	drs
Liaoning	0.683	0.745	0.917	drs
Jilin	0.591	0.660	0.896	drs
Heilongjiang	0.655	0.713	0.920	drs
Shanghai	1.000	1.000	1.000	–
Jiangsu	0.529	0.530	0.999	–
Zhejiang	0.482	0.496	0.970	irs
Anhui	0.471	0.480	0.981	drs
Fujian	0.531	0.536	0.992	drs
Jiangxi	0.475	0.485	0.978	drs
Shandong	0.537	0.552	0.972	drs
Henan	0.49	0.496	0.989	drs
Hubei	0.502	0.507	0.989	drs
Hunan	0.517	0.529	0.978	drs
Guangdong	0.599	1.000	0.599	drs
Guangxi	0.537	0.629	0.854	drs
Hainan	0.542	0.584	0.928	drs
Chongqing	0.580	0.598	0.970	drs
Sichuan	0.541	0.570	0.950	drs
Guizhou	0.579	0.677	0.855	drs
Yunnan	0.627	0.714	0.878	drs
Xizang	1.000	1.000	1.000	–
Shaanxi	0.600	0.637	0.942	drs
Gansu	0.601	0.715	0.841	drs
Qinghai	0.846	1.000	0.846	irs
Ningxia	0.566	0.892	0.635	irs
Xinjiang	0.769	0.935	0.823	drs
Median	0.608	0.675	0.914	–

Source The Author

Table 8.11 Full efficiency productivity index of express delivery network resources in 31 provinces of China in 2013

Provinces	Comprehensive technical efficiency	Pure technical efficiency	Scale efficiency	Returns to scale
Beijing	0.435	0.512	0.848	irs
Tianjin	0.729	0.733	0.994	drs
Hebei	0.513	0.925	0.554	irs
Shanxi	0.285	0.690	0.413	irs
Neimenggu	0.820	0.856	0.959	irs
Liaoning	0.728	0.734	0.992	irs
Jilin	0.690	0.701	0.984	irs
Heilongjiang	0.675	0.692	0.975	irs
Shanghai	1.000	1.000	1.000	–
Jiangsu	0.536	0.542	0.989	irs
Zhejiang	0.469	0.536	0.875	irs
Anhui	0.514	0.515	0.997	irs
Fujian	0.510	0.667	0.764	irs
Jiangxi	0.486	0.496	0.980	irs
Shandong	0.640	0.797	0.804	irs
Henan	0.499	0.503	0.992	irs
Hubei	0.477	0.486	0.982	irs
Hunan	0.465	0.472	0.986	irs
Guangdong	1.000	1.000	1.000	–
Guangxi	0.582	0.589	0.988	drs
Hainan	0.431	0.525	0.821	irs
Chongqing	0.474	0.602	0.787	irs
Sichuan	0.459	0.468	0.980	irs
Guizhou	0.719	0.732	0.982	irs
Yunnan	0.560	0.575	0.973	irs
Xizang	1.000	1.000	1.000	–
Shaanxi	0.515	0.928	0.555	irs
Gansu	0.796	0.834	0.954	irs
Qinghai	0.829	1.000	0.829	irs
Ningxia	0.887	1.000	0.887	irs
Xinjiang	0.639	1.000	0.639	irs
Median	0.625	0.713	0.887	–

Source The Author

Table 8.12 Full efficiency productivity index of express delivery network resources in 31 provinces of China in 2014

Provinces	Comprehensive technical efficiency	Pure technical efficiency	Scale efficiency	Returns to scale
Beijing	0.472	0.483	0.977	irs
Tianjin	0.702	0.708	0.991	drs
Hebei	0.425	0.425	1.000	–
Shanxi	0.398	0.428	0.930	irs
Neimenggu	0.756	0.811	0.932	drs
Liaoning	0.615	0.633	0.972	drs
Jilin	0.669	0.680	0.984	drs
Heilongjiang	0.596	0.610	0.976	drs
Shanghai	1.000	1.000	1.000	–
Jiangsu	0.480	0.481	0.998	drs
Zhejiang	0.471	0.480	0.982	irs
Anhui	0.430	0.432	0.995	irs
Fujian	0.439	0.439	0.999	–
Jiangxi	0.401	0.414	0.969	irs
Shandong	0.525	0.530	0.990	drs
Henan	0.488	0.488	0.999	drs
Hubei	0.440	0.440	1.000	–
Hunan	0.404	0.404	0.998	drs
Guangdong	0.488	1.000	0.488	drs
Guangxi	0.576	0.598	0.964	drs
Hainan	0.620	0.624	0.994	drs
Chongqing	0.503	0.505	0.995	drs
Sichuan	0.443	0.446	0.994	drs
Guizhou	0.711	0.719	0.989	drs
Yunnan	0.591	0.619	0.956	drs
Xizang	1.000	1.000	1.000	–
Shaanxi	0.460	0.478	0.963	irs
Gansu	0.585	0.635	0.920	drs
Qinghai	0.742	0.835	0.888	irs
Ningxia	0.773	1.000	0.773	irs
Xinjiang	0.586	0.620	0.945	drs
Median	0.574	0.612	0.954	–

Source The Author

Table 8.13 Full efficiency productivity index of express delivery network resources in 31 provinces of China in 2015

Provinces	Comprehensive technical efficiency	Pure technical efficiency	Scale efficiency	Returns to scale
Beijing	0.508	0.567	0.897	–
Tianjin	0.636	0.745	0.854	irs
Hebei	0.383	0.385	0.994	irs
Shanxi	0.497	0.508	0.977	irs
Neimenggu	0.823	0.844	0.976	drs
Liaoning	0.596	0.600	0.994	drs
Jilin	0.700	0.700	1.000	drs
Heilongjiang	0.627	0.630	0.996	–
Shanghai	1.000	1.000	1.000	–
Jiangsu	0.476	0.477	0.998	irs
Zhejiang	0.555	0.562	0.986	irs
Anhui	0.432	0.434	0.995	irs
Fujian	0.426	0.427	0.998	irs
Jiangxi	0.441	0.451	0.979	irs
Shandong	0.494	0.495	0.998	–
Henan	0.459	0.461	0.997	irs
Hubei	0.439	0.440	0.997	irs
Hunan	0.398	0.398	1.000	irs
Guangdong	0.792	1.000	0.792	drs
Guangxi	0.640	0.647	0.989	drs
Hainan	0.783	0.788	0.994	drs
Chongqing	0.520	0.521	0.997	–
Sichuan	0.482	0.483	0.998	–
Guizhou	0.698	0.699	0.998	–
Yunnan	0.665	0.675	0.985	drs
Xizang	1.000	1.000	1.000	–
Shaanxi	0.499	0.500	0.997	–
Gansu	0.731	0.754	0.970	drs
Qinghai	0.909	1.000	0.909	–
Ningxia	0.809	1.000	0.809	irs
Xinjiang	0.671	0.680	0.986	drs
Median	0.616	0.641	0.970	–

Source The Author

Table 8.14 Full efficiency productivity index of express delivery network resources in 31 provinces of China in 2016

Provinces	Comprehensive technical efficiency	Pure technical efficiency	Scale efficiency	Returns to scale
Beijing	0.480	0.480	1.000	–
Tianjin	0.568	0.789	0.719	irs
Hebei	0.382	0.493	0.775	irs
Shanxi	0.434	0.589	0.737	irs
Neimenggu	0.791	0.798	0.991	drs
Liaoning	0.512	0.512	0.998	drs
Jilin	0.661	0.662	0.999	drs
Heilongjiang	0.557	0.558	0.999	–
Shanghai	1.000	1.000	1.000	–
Jiangsu	0.546	0.653	0.836	irs
Zhejiang	0.487	0.506	0.963	irs
Anhui	0.380	1.000	0.380	irs
Fujian	0.383	0.387	0.989	irs
Jiangxi	0.395	0.641	0.616	irs
Shandong	0.422	0.423	0.999	–
Henan	0.413	0.487	0.847	irs
Hubei	0.413	0.422	0.979	irs
Hunan	0.389	0.645	0.604	irs
Guangdong	0.649	1.000	0.649	drs
Guangxi	0.542	0.543	0.997	drs
Hainan	0.750	0.751	0.999	drs
Chongqing	0.503	0.503	1.000	–
Sichuan	0.441	0.441	1.000	–
Guizhou	0.707	0.708	1.000	–
Yunnan	0.605	0.607	0.996	drs
Xizang	1.000	1.000	1.000	–
Shaanxi	0.453	0.453	0.999	–
Gansu	0.746	0.752	0.992	drs
Qinghai	0.998	0.998	1.000	–
Ningxia	0.659	1.000	0.659	irs
Xinjiang	0.727	0.731	0.994	drs
Median	0.580	0.662	0.894	–

Source The Author

Table 8.15 Full efficiency productivity index of express delivery network resources in 31 provinces of China in 2017

Provinces	Comprehensive technical efficiency	Pure technical efficiency	Scale efficiency	Returns to scale
Beijing	1.000	1.000	1.000	–
Tianjin	0.672	1.000	0.672	irs
Hebei	0.379	0.456	0.832	irs
Shanxi	0.438	0.673	0.651	irs
Neimenggu	0.735	0.764	0.963	drs
Liaoning	0.468	0.471	0.993	drs
Jilin	0.606	0.612	0.990	drs
Heilongjiang	0.544	0.547	0.994	drs
Shanghai	1.000	1.000	1.000	–
Jiangsu	0.494	0.511	0.967	irs
Zhejiang	0.464	0.466	0.997	drs
Anhui	0.370	0.383	0.969	irs
Fujian	0.349	0.397	0.879	irs
Jiangxi	0.402	0.698	0.575	irs
Shandong	0.401	0.403	0.996	drs
Henan	0.384	0.386	0.995	drs
Hubei	0.420	0.420	1.000	–
Hunan	0.387	0.401	0.965	irs
Guangdong	0.811	1.000	0.811	drs
Guangxi	0.497	0.502	0.990	drs
Hainan	0.742	0.759	0.979	irs
Chongqing	0.483	0.497	0.972	irs
Sichuan	0.410	0.411	0.998	drs
Guizhou	0.692	0.697	0.992	drs
Yunnan	0.549	0.560	0.982	drs
Xizang	1.000	1.000	1.000	–
Shaanxi	0.438	0.438	1.000	–
Gansu	0.691	0.714	0.968	drs
Qinghai	0.874	1.000	0.874	irs
Ningxia	0.629	1.000	0.629	irs
Xinjiang	0.710	0.733	0.968	drs
Median	0.582	0.642	0.923	–

Source The Author

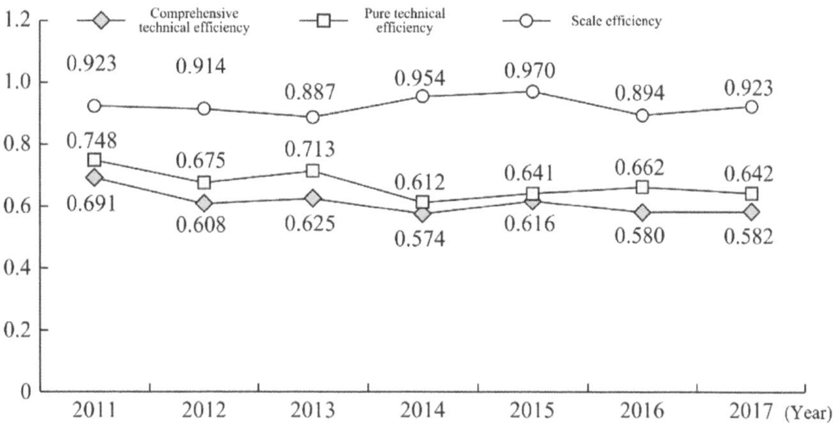

Fig. 8.4 Overall average variation in initial efficiency of the express delivery network. *Source* The Author

8.4.1.1 Analysis of Comprehensive Technical Efficiency

Excluding external environmental factors and random errors, the average level of comprehensive technical efficiency of express delivery network operations in 31 provinces in China was 0.611 over the sample period. Shanghai and Xizang had the highest comprehensive technical efficiency in China, reaching 1. Consequently, the operational efficiency of express delivery networks in these two provinces was at the relatively effective frontier, meaning that they achieved an ideal state with minimal input in the case of constant output. Provided that overall technical efficiency is unaffected, Shanghai and Xizang can exploit and amalgamate current resources so that input resources of the express delivery network can be efficiently transformed into output resources. Moreover, Qinghai, Neimenggu, Ningxia and Gansu have increased comprehensive technical efficiency, with the mean values of comprehensive efficiency of the sample data being 0.872, 0.742, 0.719, and 0.690, respectively. Conversely, the mean values of comprehensive efficiency of 19 provinces, including Beijing, Tianjin, Hebei, Shanxi, Jiangsu, Zhejiang, Anhui, and Fujian, is lower than the average, and the level of comprehensive technical efficiency in these regions is comparatively low.

8.4.1.2 Analysis of Pure Technical Efficiency

The average pure technical efficiency of express delivery network operations across the 31 provinces in China is relatively low, at 0.670, with slight fluctuations. Of note, the pure technical efficiency values of express delivery network operations in Shanghai, Guangdong, and Xizang were all 1 across the sample period, indicating that they are within the effective range of pure technical efficiency and that their express delivery networks can operate effectively under particular scale conditions. Qinghai, Ningxia, Xinjiang, Neimenggu and Gansu have relatively high pure technical efficiency values, with values of 0.972, 0.927, 0.820, 0.794, and 0.736, respectively. With the exception of 2011 and 2012, Ningxia's pure technical efficiency values were all 1. These five provinces are able to effectively manage their input resources and reach a certain level of efficiency. Beijing, Hebei, Fujian, Shandong, Hubei, Hunan, Sichuan and Shaanxi, however, all have efficiency levels below the average, with Hubei's average pure technical efficiency value being 0.473. This decreasing trend is due to the relatively high redundancy of the management organization and human resources of the express delivery network, as well as the high non-operating costs of the express delivery network.

8.4.1.3 Analysis of Scale Efficiency

Between 2011 and 2017, the scale efficiency of express delivery network operations in 31 provinces was relatively high, reaching 0.924. Shanghai and Xizang had values of 1 each year, which are located at the effective frontier of scale efficiency. This indicates that the express delivery networks in these two provinces are able to effectively utilize the scale advantage, optimize resource allocation, and benefit from economies of scale. Furthermore, Hubei, Sichuan, Liaoning, Jilin, Zhejiang, Fujian, Shandong, Chongqing, Guangxi, Guizhou, Beijing, Hainan, Yunnan and Neimenggu all demonstrate higher scale efficiency values for express delivery network operations than the average, reaching a value of more than 0.930. In contrast, the scale efficiency values of express delivery network operations in Ningxia and Guangdong are lower than the average, suggesting that the express delivery networks in these two provinces are not making effective use of scale advantages, thus resulting in lower scale efficiency values.

8.4.1.4 Analysis of Returns to Scale

In 2011, 21 provinces had decreasing returns to scale, 3 had constant returns to scale, and 7 had increasing returns. In 2012, 23 provinces showed a downward trend in returns to scale, 3 provinces showed constant returns to scale, and 5 provinces showed an upward trend. In 2013, 2 provinces had a downward trend in returns to scale, 3 provinces showed constant returns to scale, and 26 provinces had an upward trend. In 2014, 18 provinces exhibited a downward trend in returns to scale, 5 provinces showed constant returns to scale, and 8 provinces had an upward trend. In 2015, 9 provinces had a downward trend in returns to scale, 10 provinces showed constant returns to scale, and 12 provinces had an upward trend. Between 2011 and 2017, 31 provinces in China experienced reduced returns to scale for express delivery network operations, reducing from 21 in 2011 to 14 in 2017, while provinces with increasing returns to scale increased from 7 in 2011 to 12 in 2017. Consequently, provinces with reduced or static returns to scale throughout the sample period should reasonably limit the magnitude of the express delivery network and adjust the rate of expansion prudently instead of expanding their scale of operations without thought. Advanced transportation equipment and research and development technology should be implemented to enhance the operation level. Conversely, to improve the operational efficacy of express delivery networks, provinces with growing returns to scale should suitably extend their operating scale. Due to the environmental factors faced by each province, it is necessary to consider the effects of the external environment when evaluating the operational efficiency of express delivery networks in different provinces. Failure to do so may lead to results that are different from the reality. Therefore, it is necessary to adopt appropriate methods to eliminate the effects of the external environment and random errors on the operational efficiency itself.

Generally, the input–output structure of the express delivery industry in most regions is not sufficiently reasonable, necessitating further optimization, such as in terms of the input ratio of resources, resource allocation, technological means, management level, and output enhancement. With regard to resource investment, the conversion rate of the volume of express delivery business is fairly favorable, and the conversion to output has been realized in the majority of regions. However, express delivery outlets and total postal route length have yet to maximize conversion efficiency, and many regions experience serious resource redundancy. In addition to the issues of scale efficiency and economies of scale in the regional express delivery industry, further analysis is provided. Zhejiang should increase and improve the infrastructure of express delivery outlets in order to improve industrial operational efficiency and income. Hebei should establish more express delivery outlets and improve the postal route infrastructure. Shanxi, Fujian, Jiangxi, Henan, Hubei, and Hunan should increase the volume of express delivery business and, simultaneously, the number of express delivery postal routes. Beijing, Shaanxi, Neimenggu, Shandong, Liaoning, Jilin, Yunnan, Guangxi, Heilongjiang, Qinghai, Chongqing, Sichuan, Guizhou, Hainan, Gansu, and Xinjiang need to reduce their investment in express delivery outlets and postal route infrastructure in order to reduce investment costs. Anhui, Guangdong, and Ningxia should expand their industrial scale in suitable regions with reasonable resource investment structures.

Overall, the operational efficiency of the express delivery industry is low and the issues vary across regions. In general, only a few regions have higher efficiency. Most regions have inefficient industrial structures, and the utilization of express delivery network resources is not high. The causes of DEA ineffectiveness in regions that have not achieved it are various and the same factor contributes to the overall number of ineffectiveness. Thus, through the DEA effectiveness analysis, the authors suggest that relevant enterprises and industries should focus more on balance between elements when expanding their business, rather than the absolute number. When developing and utilizing express delivery resources, all resources should be allocated in a Pareto optimal manner in order to improve the utilization efficiency.

8.4.2 SFA Regression Analysis Results

In the initial stage, the input slack variables were obtained. In the second stage, random frontier regression analysis was conducted using Frontier 4.1 software to reduce the effects of environmental factors and random errors on the DMU. The dependent variable of the stochastic frontier regression analysis was the input slack variable. There were a total of three dependent variables. The three input slack variables represented the investment difference in express delivery outlets, the total length of postal routes, and the volume of express delivery business. Three explanatory variables were included, namely environmental variables (infrastructure construction, the number of employees in the express delivery industry, and the average number of customers in each outlet). The results of random frontier regression are presented in Tables 8.16, 8.17, and 8.18, where * denotes the regression passed the 10% significance test, ** denotes the regression passed the 5% significance test, and *** denotes the regression passed the 1% significance test.

Table 8.16 Second-stage regression analysis of input slack of express delivery outlets

	Coefficient	Standard error	T-ratio
Constant term	1.31E+03	1.00E+00	1.31E+03
Number of employees	−2.40E−02	3.95E−02**	−6.08E−01
Infrastructure construction	2.70E+00	6.78E−01	3.98E+00***
Customers	−1.92E+03	1.00E+00	−1.92E+03
σ^2	9.94E+06	1.00E+00	9.94E+06
γ	1.00E+00	1.68E−01	2.83E−01
Log likelihood function value = −0.29E + 03***			
LR test of the one-sided error = 0.33E + 02**			

Source The Author

Table 8.17 Second-stage regression analysis of input slack of total length of postal routes

	Coefficient	Standard Error	T-ratio
Constant term	−9.68E+04	1.00E+00	−9.68E+04
Number of employees	−2.24E+00	3.17E−01	−7.09E+00
Infrastructure construction	8.16E+01	9.02E−01	9.05E+01
Customers	2.09E+04	1.00E+00	2.09E+04
σ^2	2.10E+10	1.00E+00	2.10E+10***
γ	1.00E+00	9.54E−06	1.05E+05
Log-likelihood function value = −0.39E+03***			
LR test of the one-sided error = 0.81E+01			

Source The Author

Table 8.18 Second-stage regression analysis of input slack of the volume of express delivery business

	Coefficient	Standard error	T-ratio
Constant term	−4.09E+04	1.00E+00	−4.09E+04***
Number of employees	−4.09E−02	6.42E−02	−6.37E−01
Infrastructure construction	2.84E+01	1.94E+00	1.46E+01
Customers	−1.35E+05	1.00E+00	−1.35E+05
σ^2	1.70E+10	1.00E+00	1.70E+10
γ	1.00E+00	8.02E−06	1.25E+05
Log-likelihood function value = −3.89E+02**			
LR test of the one-sided error = 1.24E+01			

Source The Author

As shown in Tables 8.16, 8.17 and 8.18, most of the three environmental variables were found to have passed the significance test, indicating that external environmental factors significantly influence the investment redundancy of the express delivery network. The γ values of the three input slack variables, namely the network slack variable, the total length of postal route slack variable, and the slack variable in the volume of express delivery business, were all 1.00, and all were found to have passed the significance test. The results demonstrate that, in the development of the express delivery network, random errors have little effect on input slack variables. Nevertheless, the management level still significantly affects the operational efficiency of the express delivery network. Therefore, the use of the SFA regression analysis model is essential in order to mitigate the impacts of environmental factors and random errors on efficiency value.

The positive and negative values of the regression coefficients signify different implications when assessing the three regression models above. If the regression coefficient is positive, an increase in external environmental variables will result in a waste of resources or a decrease in output. Conversely, if the regression coefficient is negative, the opposite outcome is expected. The impact of the three environmental variables is thus analyzed as follows.

1. The number of employees

   The number of employees in the express delivery industry has a significant positive correlation with the slack variable, which is statistically significant at the 1% significance level. This indicates that an increase in the number of employees in the express delivery industry will result in the waste of investments. Thus, blindly increasing the number of employees in the express delivery industry will lead to an increase in input factors, resulting in the inefficient utilization of resources and detrimentally affecting the environment.
2. Infrastructure construction

   The regression coefficients of infrastructure construction on slack variables are all positive and have all passed the 1% significance test. This suggests that expanding the scale of infrastructure construction will lead to an increase in slack variables, demand for labor, and length of express delivery network; consequently, these factors are not conducive to rational resource use and may result in excessive investments. Likewise, this reveals that infrastructure construction is an unfavorable environmental variable. The expansion of infrastructure may lead to the misallocation of capital and human resources, thereby reducing operating efficiency.

3. The average number of customers in each express delivery outlet

   The environmental variables had a negative correlation with the slack variables. As the number of customers increases, full utilization of the resources of the express delivery network will enhance its operational efficiency.

8.4.3 Adjusted Input Quantity in BCC Model Calculations

According to the SFA regression model established in the second stage, 31 provinces in China are considered to be in the same environment with the original input data adjusted. Using the BBC model, efficiency values were calculated based on the adjusted input values and output values of the first stage. It can be seen that the efficiency values of the 31 provinces in China have significantly changed compared to those of the first stage. The calculation results are presented in Table 8.19.

Table 8.20 in this chapter utilizes the average efficiency value of express delivery networks in 31 provinces as the evaluation value of efficiency, and the operational efficiency of express delivery networks is studied accordingly. The average efficiency value from 2011 to 2017 is lower than pre-adjustment to different extents, as demonstrated in Fig. 8.5.

From an average pure technical efficiency perspective, the adjusted pure technical efficiency has improved, suggesting that the express delivery network's operating efficiency is influenced by environmental factors and random errors. Consequently, the express delivery network should modify the operating environment, which is beneficial to enhancing the overall efficiency of the express delivery network.

From a perspective of scale efficiency, the adjusted scale efficiency is lower than previously observed, suggesting that environmental factors and random errors are significantly influencing scale efficiency. Weak infrastructure construction, irregular management of postal employees, and an uneven distribution of the number of people served at each outlet may be contributing to this situation. Thus, it is necessary to reinforce the management of the operational network.

The average comprehensive technical efficiency of express delivery networks in 31 provinces in China decreased by 0.388, 0.234, 0.382, 0.329, 0.335, 0.290, and 0.101 from 2011 to 2017. This indicates that the overall operational efficiency of China's express delivery networks is low and the decrease is evident, which necessitates further improvement. From 2011 to 2017, the average pure technical efficiency increased by 0.085, 0.294, 0.156, 0.236, 0.204, 0.173, and 0.115, while the average scale efficiency decreased by 0.572, 0.527, 0.608, 0.657, 0.629, 0.532, and 0.349,

Table 8.19 Average efficiency of express delivery network resources for 31 provinces of China in the third stage

Provinces	Comprehensive technical efficiency	Pure technical efficiency	Scale efficiency	Returns to scale
Beijing	0.489	0.866	0.564	irs
Tianjin	0.377	0.850	0.444	irs
Hebei	0.337	0.793	0.424	irs
Shanxi	0.201	0.710	0.283	irs
Neimenggu	0.269	0.772	0.349	irs
Liaoning	0.368	0.605	0.609	irs
Jilin	0.298	0.768	0.388	irs
Heilongjiang	0.291	0.713	0.408	irs
Shanghai	1.000	1.000	1.000	–
Jiangsu	0.474	0.675	0.702	irs
Zhejiang	0.701	0.804	0.871	irs
Anhui	0.331	0.661	0.501	irs
Fujian	0.415	0.880	0.472	irs
Jiangxi	0.279	0.671	0.416	irs
Shandong	0.472	0.707	0.667	irs
Henan	0.374	0.598	0.624	irs
Hubei	0.413	0.671	0.615	irs
Hunan	0.321	0.675	0.476	irs
Guangdong	0.853	1.000	0.853	drs
Guangxi	0.348	0.741	0.469	irs
Hainan	0.138	0.851	0.162	irs
Chongqing	0.367	0.804	0.456	irs
Sichuan	0.474	0.630	0.753	irs
Guizhou	0.434	1.000	0.434	irs
Yunnan	0.594	1.000	0.594	irs
Xizang	0.062	1.000	0.062	irs
Shaanxi	0.351	0.760	0.462	irs
Gansu	0.196	0.709	0.276	irs
Qinghai	0.055	0.962	0.057	irs
Ningxia	0.107	1.000	0.107	irs
Xinjiang	0.314	0.895	0.351	irs
Median	0.377	0.799	0.479	

Source The Author

Table 8.20 Average efficiency values before and after adjustment from the 31 provinces in China from 2011 to 2017

	2011	2012	2013	2014	2015	2016	2017	Mean
The comprehensive technical efficiency before adjustment	0.691	0.608	0.625	0.574	0.616	0.580	0.582	0.583
The adjusted comprehensive technical efficiency	0.303	0.374	0.243	0.245	0.281	0.290	0.481	0.377
The pure technical efficiency before adjustment	0.748	0.675	0.713	0.612	0.641	0.662	0.642	0.630
The adjusted pure technical efficiency	0.833	0.969	0.869	0.848	0.845	0.835	0.757	0.799
The scale efficiency before adjustment	0.923	0.914	0.887	0.954	0.970	0.894	0.923	0.939
The adjusted scale efficiency	0.351	0.387	0.279	0.297	0.341	0.362	0.574	0.479

Source The Author

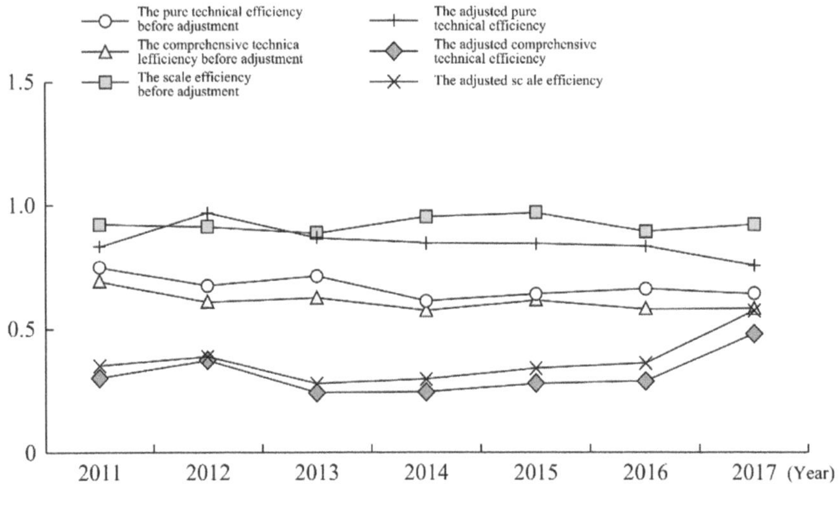

Fig. 8.5 Average efficiency values in the 31 provinces of China from 2011 to 2017 before and after adjustment. *Source* The Author

respectively. This demonstrates that the low operational efficiency of China's express delivery network is mainly reflected in the low scale efficiency, which is due to the low utilization rate of the express delivery network, thus constraining the expansion of the express delivery network.

8.5 Summary

Using a seven-year sample data set of express delivery networks in 31 provinces of China, this chapter applies a three-stage DEA method to analyze the operational efficiency of China's express delivery industry. The operational efficiency of the express delivery industry in 2017 was evaluated from four perspectives: comprehensive technical efficiency, pure technical efficiency, scale efficiency, and returns to scale. The results indicate that the overall operational efficiency of the express delivery industry is low and that management levels significantly affect the operational efficiency of express delivery networks. In addition, an SFA model was used to select the panel data from 2011 to 2017, eliminating environmental variables and random errors, and making adjustments to input–output variables. The adjusted input–output variables were then calculated by DEA, making the evaluation results more reasonable and accurate.

References

Charnes, A., Cooper, W. W., & Rhodes, E. (1978). Measuring the efficiency of decision making units. *European Journal of Operational Research, 2*(06), 429–444.

Chen, Y., Du, J., Sherman, H. D., & Zhu, J. (2010). DEA model with shared resources and efficiency decomposition. *European Journal of Operational Research, 207*(01), 339–349.

Du, J., & Huo, J. Z. (2014). DEA-based evaluation on city innovation in China. *Chinese Journal of Management Science, 22*(06), 85–93.

Liu, J. B., & Gao, P. (2010). Study of the GEM listed companies' performance evaluation in China based on BBC model. *China Soft Science* (S1), 263–270.

Wang, M.Q., Liang, L., & Li, Y. J. (2009). Fuzzy super-efficiency DEA model. *Chinese Journal of Management Science, 17*(02), 117–124.

Zhang, A. F., & Che, X. H. (2011). An analysis on the efficiency of Guangzhou technology innovation and its influence factors based on DEA. *Science and Technology Management Research, 31*(24), 20–23.

Chapter 9
Management Mode of Express Delivery Networks

Currently, the management mode of the express delivery network can be classified into direct operation, franchise or agency, and self-operated modes of operation. However, there are associated issues that impede the growth of the express delivery industry and the customer experience. Thus, reforming the management mode of the express delivery network could promote the development of the express delivery industry to some extent.

9.1 Management Mode of Express Delivery Networks

Express delivery network is an extended component of the transportation network, which is a system connected by transit centers and transportation routes according to specific principles and methods (Gao & Tao, 2013). Express enterprises and express delivery networks are interdependent, with express delivery networks being the primary transportation system for express enterprises to transport goods. It is composed of physical, business, and control layers. Chapter 7 provides a detailed explanation of the relationship between each layer of express delivery networks. The physical layer serves as the foundation of the express delivery network system, while the business layer constrains the system. The control layer supervises, controls, and coordinates with the business layer to calculate the optimal transportation routes and directions of express shipments to ensure efficient, effective transmission and smooth flow throughout the network (Jiang, 2018). It is known that express delivery network is characterized by integrity. The system ensures the efficiency of express transportation by relying on the overall functioning of the network and the interdependent support of multiple layers. The operations of express delivery enterprises are contingent upon the normal functioning of express delivery networks; the prerequisite for express delivery enterprises to make a profit is the normal functioning of

P. Li and J. Wu, *Research on Innovation of Express Delivery Network Management*,
https://doi.org/10.1007/978-981-99-8982-9_9

express delivery networks. Therefore, the management model of express delivery is dependent on the model of express delivery networks.

Management mode is a management behavior structure constructed in accordance with management concepts, methods, models, systems, tools, and procedures. Enterprise management mode is centered around knowledge management, with opportunity management at its core. Management mode is also referred to as a management model, which is a type of management divided by certain characteristics, and could be seen as a particular form of management distinguished by specific features. This model is intended to make the organization more dynamic and to provide democratic and transparent management for employees.

The management mode of express delivery networks is a systematic, efficient, and relatively complete system formed through the combination of express delivery networks and management modes, thereby contributing to the development of the entire express delivery industry. As demonstrated in Chapter 3, through the development and current status of the express delivery industry, it can be seen that the majority of China's delivery outlets are located in Eastern China, Northern China, Central and Southern China. Among them, China Post's express delivery network is the most widely spread. According to Table 9.1, the characteristics of express delivery enterprises are divided by ownership. Additionally, based on Table 9.1 and the development of both China's and foreign delivery industries in Chapter 3, the management mode of express delivery networks can be further categorized into alliance mode, direct-operated mode, franchise or agency mode, and self-operated mode.

9.1.1 Direct-Operated Mode

9.1.1.1 Overview of Direct-Operated Mode

A corporate headquarter adopts a vertical management approach to directly invest, manage, and operate all outlets, which is known as the direct-operated mode. Enterprises expand their own strength and scale through annexation, mergers, and other means. Under this mode, the arrangement is much more standardized; the headquarters allocate resources such as staff and transportation vehicles to each district of the express enterprise in accordance with their respective business volumes. This mode is characterized by the corporate headquarters and branch office belonging to the same legal entity, and the interests of the two being fundamentally aligned. However, companies with this management model will increase operating costs, reduce management flexibility, and impede the development of their network. Table 9.2 in the text displays the express enterprises and their scales in the direct-operated mode in China.

SF Express implements a direct operation management model wherein the headquarter has the authority to manage and operate the enterprise, and the branches of the direct chain implement standardized enterprise management. To control all

Table 9.1 Category, representative enterprise and characteristics of express delivery enterprise

Enterprise category	Representative enterprise	Characteristics
State-owned express delivery	Express Mail Service (EMS)	It has policy advantages, developed transportation network, high price, diversified service types and complete transportation equipment, and it has obvious advantages over transportation speed and quality, so it occupies the middle and high-end share of the express delivery market
Private express delivery	SF Express, STO Express, YTO Express, ZTO Express, JD Logistics	These enterprises develop quickly, has flexible price, and compete highly. Relying on its own strong strength, SF Express has occupied a large share of the domestic middle and high-end market. In recent years, JD Logistics also has its own express line, which has significantly improved in quality and transportation time, and it has slowly entered the high-end market. And other private express delivery enterprises (STO, ZTO, YTO Express) occupy the low-end market
Foreign-funded express delivery	UPS, FedEx, DHL, Japan Post	They have strong strength, standard operation, but relatively high price. Based on its own advantages, it mainly targets the international express delivery market in China

Source The Author

Table 9.2 Express delivery in direct management mode

Enterprise name	Size	Organizational form
SF Express	It has own 41 freighters, 16 leased freighters, 62 high-speed rail lines and 96 ordinary rail lines. It cooperates with 673 property management company outlets with nearly 30,000 cooperative agencies	Direct management model
EMS	There are 33 freighters, and the scale of autonomous aviation continues to expand. It opened high-speed rail mail routes to achieve "same-day delivery" and "next-day delivery"	Direct management model

Data Source Express Market Regulatory Report in 2017

branch points, SF Express implements top-down management, which includes standardized and unified management such as departmental staff training, operational performance assessment, punishment measures, pricing of express packages, and the flow of express packages (Li, 2018). Additionally, all department employees of

SF Express are uniformly mobilized by the headquarter's human resources department. For instance, the headquarter provides and develops customers for dispatchers in order to ensure the fairness of resource allocation.

9.1.1.2 Advantages of Direct-Operated Mode

The primary benefit of this approach is that it unifies operations and management, defines service prices and target customers, implements policies and management effectively, establishes a unified corporate image, and sets consistent goals for corporate development. This is evidenced through the four unifications that express enterprises are able to achieve: funding, business strategies, personnel management, and the overall resources of the enterprise, thus enabling the full use of the advantages of integrity. Additionally, this mode entails that the express company bears all benefits and risks generated by each outlet within the express delivery network.

9.1.1.3 Disadvantages of Direct-Operated Mode

The major disadvantage of this mode is that enterprises need to invest considerable funds in order to set up their transport sector, rent factories, equipment, and transportation tools, as well as expand the coverage area of outlets. Furthermore, strict monitoring of the number of outlets in various regions leads to a relatively small express service network coverage, and the expansion of these services is relatively slow. The investment cycle for express operations is long, causing the management cost to be relatively high. Brand building, team building, business development, and culture building take a long time, and each outlet has little autonomy in terms of operating rights, resulting in employees not being fully motivated due to incompact interest relationships.

9.1.2 Franchise or Agency Mode

9.1.2.1 Overview of Franchise Mode

The franchise mode involves the headquarters establishing a total number of franchises by region, which are then divided into smaller franchises, with the most basic level being an express outlet. The headquarters provides the company's trademarks, patents, etc. to be used by the franchises in exchange for payment in the form of contracts, and also provides guidance and training, as well as access to the express delivery network system. The management and training of franchising involves being familiar with the organizational structure and other necessary resources to help the franchises quickly understand the business process. Its main features include the legal status of the headquarters and the franchises being equal, as both are independent

Table 9.3 Express partnership mode

Company name	Size	Organizational form
ZTO Express	It has more than 2,000 long-distance transportation routes, over 3,600 trucks, about 1,800 high-capacity trucks, and 1,200 outsourced transporters	Partnership model
STO Express	There are 1,830 main transportation routes, 3485 trunk transport vehicles, 646 new trailers and 550 tractors	Partnership model
YTO Express	It's share of bilateral transport vehicles continues to increase, increases its own trunk transport vehicles, and improves the trunk line system	Partnership model
Yunda Express	It has over 4,300 regular land trunk transportation routes and cooperates with more than 30 air freight forwarding companies	Partnership model

Data source Express Market Regulatory Report in 2017

legal persons, which can lead to the problem of divergent interests. Table 9.3 shows the express delivery enterprises and their scales when using the franchise mode.

9.1.2.2 Advantages of Franchise Mode

The advantages of this mode include low operating costs, high management flexibility and rapid network development. Franchisees pay franchise fees to obtain operation rights, which means that the headquarters can quickly expand and occupy the express delivery market without the need for a great deal of investment (Yu & Zhang, 2014). Furthermore, franchisees have the freedom to recruit employees, set pricing, and devise business strategies. By investing some funds, franchisees can acquire exclusive operation and management rights of a certain express delivery network in a certain area in a relatively short period of time, enabling rapid business development. This mode also allows the sharing of the brand and network of the headquarters, thus providing relatively low-risk and high-efficiency entrepreneurship. Moreover, the corporate headquarters and franchisees are independent legal persons, making the management cost low, and they do not share the benefits of the operation.

9.1.2.3 Disadvantages of Franchise Mode

Despite the many advantages of franchise mode, there are certain shortcomings and loopholes as well. For example, franchisees experience great difficulty in management, pricing is chaotic, operation and service quality vary significantly, and execution of different outlets is often inconsistent. Additionally, corporate image and corporate goals vary from one outlet to another, leading to confusion or even lack of target customers' consumption behavior.

9.1.3 Self-Operated Mode

9.1.3.1 Overview of Self-Operated Mode

In self-operated mode, enterprises are able to independently establish and manage their own distribution systems, thereby achieving internal and external goods distribution. This mode is beneficial for the integration of enterprise supply, production and sales, and is highly systematic. There are distinctions between self-operated and direct-operated modes. The direct-operated mode involves top-down management by corporate headquarters and the implementation of unified operations, investments, and management of each branch point. The self-operated mode is also directly managed by the headquarters, but it is a combination of e-commerce enterprises and the express delivery industry, constituting a self-operated logistics distribution mode in which the entire logistics system is part of the enterprise, with the main service goal being the enterprise's production and operation.

JD Logistics operates on a self-operated basis. In 2007, JD began constructing its own logistics system. The following year, Shanghai Yuanmai Logistics and Express Co., Ltd. was established in Shanghai. In 2009, JD invested in the construction of its largest storage center, the "East China Logistics Storage Center", in Shanghai. Since then, the self-operated logistics system of JD has gradually matured. Eight major logistics hubs have been established, with Beijing, Shanghai, Guangzhou, Chengdu, Wuhan, Shenyang, Xi'an, and Hangzhou at the center, based on regional traffic and customer consumption orders in each city. This is the largest intelligent logistics warehouse group in the field of e-commerce logistics in Asia. JD has also established distribution stations in 300 major cities, such as Tianjin, Suzhou, Hangzhou, and Nanjing. As of June 2019, JD has seven first-class logistics centers, a professional distribution team of over 55,000 people, approximately 600 warehouses, and 6,756 express outlets throughout China, which serves a total of 2,639 districts and counties (Tian, 2017). Its operational model is depicted in Fig. 9.1.

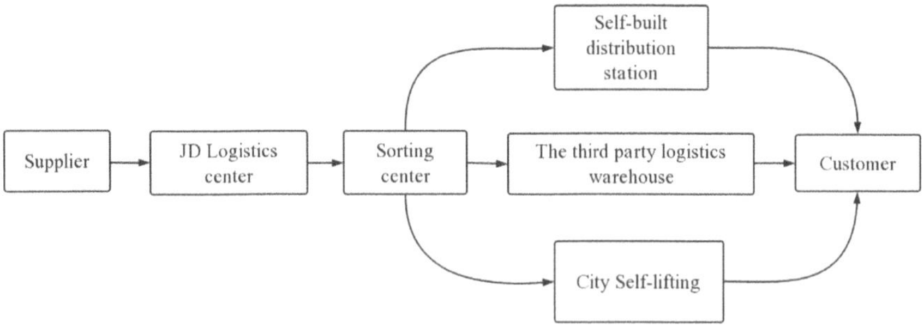

Fig. 9.1 JD Logistics operation mode *Source* The Author

9.1.3.2 Advantages of Self-Operated Mode

The logistics distribution in this mode is both timely and safe. By directly controlling logistics-related businesses, e-commerce companies are able to ensure the service quality and efficiency of the logistics and distribution process, which in turn improves user satisfaction. However, if JD and a third-party logistics have established a commercial partnership, there could be uncontrollable factors that hinder the development of logistics to some extent. By providing this service via JD's exclusive distribution team, they can protect the safety of the goods while maintaining the company's image. This also ensures the stability of the business, as well as the brand and advertising effects of self-operated logistics during special periods. During special periods such as legal holidays, the express volume increases rapidly and most express enterprises take holidays for their staff, which can lead to delays in packages and dissatisfaction among consumers. By basing their warehouses and orders on location, self-operated logistics systems can effectively reduce the pressure of over-stocking and transportation during peak periods, providing not only business stability but also a certain level of advertising and brand image for e-commerce companies. Moreover, self-operated logistics maximize customer satisfaction with regards to dispatching and services, thus improving the recognition of e-commerce enterprises and corporate brands.

9.1.3.3 Disadvantages of Self-Operated Mode

Initially, this mode has high investment and risks, making it difficult to focus on core business. On the one hand, the self-operated mode requires a significant investment in order to improve its logistics system, resulting in a greater financial burden on the enterprise. Therefore, when enterprises invest in logistics, they must also reduce their capital investment in other sectors. The inability to recoup costs within a short period of time leads to a disruption of the capital flow, thus significantly diminishing the capacity of enterprises to withstand market risks. On the other hand, the core business of self-operated logistics companies is not logistics and distribution services. As the company invests considerable time and resources into the self-built logistics business, it can hinder the core business of the company and decrease its overall flexibility. Another downside is the increased pressure on logistics and distribution. While the self-built logistics system has been recognized by consumers, ensuring the security and timeliness of logistics and distribution, the additional orders raise the demands on logistics and distribution, thus leading to greater pressure on logistics and distribution.

9.2 Factors Influencing Express Delivery Network Management Reform

In the initial part of this chapter, the operating model of express enterprises is examined in depth. Given the current rapid growth of express delivery companies, there are pressing issues requiring immediate resolution in terms of the management of the express delivery network. These mainly include the market, pricing, logistics infrastructure, selection of delivery sites, and policy.

9.2.1 Market Factors

9.2.1.1 Low Barrier to Entry in the Express Delivery Industry

The rapid proliferation of online shopping has led to an exponential rise in the number of express delivery companies. As a result of government policy support and lax capital requirements, it is easy to enter the industry. Consequently, there is a vast disparity in the qualifications of industry participants. The low investment and cost associated with express delivery services have encouraged many people to attempt to capitalize on the market. Consequently, an abundance of companies have entered the industry, leading to inevitable cutthroat competition.

9.2.1.2 Disorganization of Express Delivery

During the annual "Double 11", express delivery companies are faced with hundreds of millions of express shipments from thousands of customers that need to be delivered simultaneously. In 2017, some companies were unable to adequately address the sheer volume of packages, leading to frequent chaos and occasional violence in sorting. Imperfect relevant laws and regulations, as well as weak supervision, have caused the chaos in the express delivery industry to be repeatedly prohibited, yet it persists. In fact, the chaos of express is not limited to violent sorting, but also includes express without real name, loss of packages, difficulty in seeking compensation, delay in dispatching, bad attitude, and leakage of customers' information.

9.2.1.3 Lack of Professional Delivery Talents

The lack of professionally trained delivery personnel constitutes an important element hindering the progress of the express delivery network to a higher level. Currently, the hiring standards for deliverymen are relatively lax; typically, those who are familiar with the layout of the delivery region can become deliverymen, with no other professional qualifications being necessary. In recent times, the introduction of new delivery methods such as delivery lockers and unmanned aerial vehicles has necessitated that couriers learn and become adept at new theories, technologies and tools. As a result, professionally trained delivery personnel are in short supply.

9.2.1.4 Low Quality of Employees

The express delivery industry is labour-intensive. In order to facilitate rapid development and expansion of outlets, express delivery companies require a large number of employees, which can lead to a disregard for the necessary qualifications when selecting employees. Despite the stabilization of China's express delivery industry, the courier staff in the industry generally lack service awareness. The phenomenon of customers complaining about courier service problems, such as delays in delivery, collection of payments, illegal charges, and item damage, is increasing year by year, which will adversely affect the overall image of the enterprise and lead to unnecessary losses (Lin, 2016). Low comprehensive quality of express employees is likely to become a major issue restricting the development of the enterprise upon comprehensive analysis.

9.2.1.5 Lack of Cooperation among Express Delivery Companies

Most of the current express delivery companies in the country are engaged in single transportation, warehousing and distribution using a single profit mode and offering few value-added services; as a result, they find it difficult to resist greater industry risks. Although the emergence of Cainiao Station and Express Alliance has increased cooperation between various express delivery companies to some degree, urban distribution companies still operate independently and each of them persistently utilizes closed logistics, advanced technology, management methods, and customer resources. Consequently, a lack of communication and cooperation between companies makes it difficult to form a mutually beneficial and win–win community.

9.2.1.6 Low Degree of Humanization

According to regional differences, the business requirements vary, resulting in an uneven distribution of business development among different outlets. For example, some regions are populated by a large number of college students, while other regions are primarily business-oriented. Therefore, express delivery companies must adjust the service capabilities of outlets based on the characteristics of regional business. Additionally, changes in the external environment may necessitate the temporary deployment of couriers to meet customer demands, thereby increasing the difficulty and intensity of management and not meeting the requirements of refined management to a certain extent. Furthermore, as the agency density increases, the labor intensity of couriers also increases, potentially leading to delays in delivery and decreased customer satisfaction.

In comparison to other private logistics companies, EMS cannot provide direct delivery to customers except for certain special items; instead, packages must be dropped off at post offices, street offices, and other locations, resulting in considerable inconvenience for customers.

9.2.2 *Price Factor*

9.2.2.1 Non-uniform Pricing Standards for Express Delivery

In 2015, the National Development and Reform Commission and the State Post Bureau jointly issued the *Notice on the Relevant Issues Concerning the Releasing of Some Postal Service Fees*, which clearly defined the domestic express mail fee, postcard delivery fee, printed matter delivery fee, and "bubble counter" packages weighing less than 10 kg (with packages' weight not exceeding 167 g per cubic decimeter) and other competitive packages delivery prices as being changed from government pricing to market-regulated pricing. Postal companies were then able to independently determine the freight price structure, freight price standards and billing according to market supply and demand and competition conditions. This meant that the express freight price of most online shopping packages was determined by the express company itself, resulting in an uneven express pricing across the express delivery industry.

9.2.2.2 Low Degree of Marketization of Price Formation

Express is a competitive service industry, and its price is primarily determined by the market. It is formulated in accordance with the "floating" pricing mechanism and the market rules of discounted prices during off-season and full-prices during peak periods. However, many express delivery companies in China violate this pricing mechanism, ignoring the market principle of using price levers to adjust market demand, and repeatedly reduce express prices in order to gain a monopoly in the market. At present, e-commerce express mail holds a significant share of express items transported in China. In the e-commerce express mail market, the more packages sent, the lower the unit price. In order to gain access to the price differentials of intermediate express, some e-commerce companies take advantage of their source of goods to unreasonably reduce the express price. This has subsequently led to a lack of corporate financing funds for express delivery companies and an inability to introduce high-performance software and hardware equipment.

9.2.3 Logistics Facilities

Express delivery networks are complex systems composed of outlets and lines. Reasonable location selection and route planning can substantially lower the construction cost of express delivery networks. Currently, numerous scholars have studied and proposed methods to optimize the physical layer of express delivery networks, albeit with a low degree of application. The reasons for this are fourfold.

(1) The cost of optimizing the existing network system is high, making it unaffordable or unwilling to be afforded by some enterprises.
(2) In the process of increasing business volume, express delivery companies will not devote their resources to restructuring work. With current efficiency improving steadily, companies also lack the incentive to reform.
(3) The network optimization and reconstruction cycle is lengthy and expensive. If only the local network is optimized, the service capability of the regional network may not be significantly improved, and the service stagnation or quality deterioration during the implementation process could negatively impact the reputation of the enterprise.
(4) Some Enterprises Lack the Strategic Vision Necessary for Long-term Planning.

9.2.3.1 Slow Integration of Network Resources

Resource integration has consistently been a challenging issue faced by express delivery companies. Through analyzing the nation's policies on the express delivery industry in recent years, the Chinese government has consistently highlighted the necessity of adjusting the industrial structure, rationally integrating resources, and increasing resource utilization. Resource integration is a mandate put forward by the state for the express delivery industry, and all express delivery companies should take this as a development idea. For express delivery companies themselves, the traditional express delivery network model is no longer capable of satisfying actual requirements. Because of out-of-date information technology, manual operations are still carried out in most sorting centers.

9.2.4 Delivery Location

"The 'last mile' delivery is of great importance and is a critical element in the fulfillment of the agency process; Moreover, it is an essential link for expressing service quality in delivery. Currently, there are numerous issues faced by the 'last mile express' distribution, with customers often filing complaints in regard to it. This has become a hindrance to the growth of the express delivery industry."

9.2.4.1 Challenges in Delivery Management Coordination

The distribution of the express delivery network mainly involves the operation management department, the transportation department and the urban planning department. These departments perform their own duties with their own work areas and working principles. So it is easy to have divergence in the management of express vehicles, provisional express pickup points and express centers. The Operation Management Department is responsible for the transportation and operation of goods via roads. As a management department, it must not only implement the nation's development strategy but also adhere to the pertinent regulations of local provinces and cities. The department will look to protect the interests of urban distribution companies and reduce the limitations of the Transportation Department on shipping. The purpose of the Transportation Department is to regulate transportation, which could potentially clash with the operations of the Operation Management Department. Furthermore, the Urban Land Planning Department did not plan roads and express delivery outlets in accordance with the development of the existing express delivery network, which ignored the customer demand for express delivery to a certain degree, and failed to take into account urban delivery stations and road reservation matters.

9.2.4.2 Inaccurate Network Planning

According to the requirements for service capacity during the planning process, the construction of regional outlets should aim to ensure that agency coverage is not duplicated and as wide as possible, whilst also considering cost and management when aiming for a high population coverage rate. If this is not done, it can result in wasted resources and increased investment costs, as well as confusion in terms of management. In areas where agencies are particularly concentrated, the responsible outlets should strive to improve their service capacity, as well as increasing their dispatch resources. The selection of the site should comprehensively consider factors such as cost, areas with high agency involvement and the shortest postal route. Taking SF Express as an example, due to all its express outlets being self-constructed, the expansion speed and coverage rate of its outlets are slow. This means that some towns or remote areas in second- and third-tier cities are typically not within the scope of SF Express's services, which severely hinders its development. Even though the management and operation of SF Express are done through a direct-operated mode, the large number of employees and branches it has makes it difficult for the ideas of the superiors to be accurately conveyed to lower-level employees, thereby leading to difficulties in management as internal "small groups" form.

9.2.4.3 Increasing Traffic Burden due to Volume Distribution

The ongoing development of the express market has necessitated express delivery companies to augment their business outlets and widen their distribution areas. To satisfy the increasing express demand from customers, companies must invest more labor and material resources, which is mainly seen in the growth of delivery vehicles. Presently, the delivery tools for the last mile are predominantly three-wheeled motorcycles. However, the delivery tools are parked without any regulation, thus exacerbating the already congested traffic conditions.

9.2.4.4 A High Rate of Freight Under-Load

With the development of express delivery methods and the competition among express delivery companies, customers have an increasing demand for timeliness when it comes to express delivery services. Therefore, delivery time has become a key factor in the long-term development of express delivery companies. Traditional express delivery is unable to satisfy the demands in terms of transportation efficiency and time cost. Although the freight transport industry is shifting away from its traditional reliance on manual labour, and embracing the Internet, the empty load rate of China's freight transport was as high as 14% at the end of 2017, indicating that return vehicles often struggled to find suitable sources of goods. Express truck drivers travel up to 1,500 km a day, and the longest time interval for parking and loading of transport vehicles is 72 h; this reveals the current market's inefficient resource allocation and the wastage of resources.

9.2.5 Policy Factors

9.2.5.1 Inadequacy of Regulations on Express Delivery

The development and innovation of express delivery networks cannot be separated from the strong support of China's government and relevant departments, yet the support of relevant government departments still needs to be strengthened. In March 2018, the "Interim Regulations on Express Delivery", signed by State Council Premier Li Keqiang, came into effect on May 1, 2018. Although Article 13 of the draft explicitly states the principle of "protecting the rights of express service vehicles to pass and temporarily stop in accordance with the law, and not prohibiting express service vehicles from passing by law", there is a lack of operability, and no specific penalty method for this issue. Furthermore, this regulation has fewer contents related to the delivery of the new express delivery networks that have emerged in cities.

9.2.5.2 Weak Regulation of Prices in the Express Delivery Market

China's policies and regulations concerning express delivery services are gradually improving due to the revision, promulgation and implementation of the Postal Law of the People's Republic of China. Taking into account the current situation, the management department has formulated relevant regulations on express delivery services from various perspectives and is supervising and managing them. Nevertheless, price regulation in the express market still needs to be improved. The direct reason for the lack of price supervision is the imperfection of existing laws and regulations, which has resulted in chaos in the express market price order and a joint price increase from some express delivery companies. Additionally, the prices of each franchisee of franchise express delivery companies are not uniform, which is an internal factor in the chaos of the price system and the random quoting of couriers in the express delivery industry.

9.2.5.3 Unclear Express Delivery Network Planning

The express delivery business is currently experiencing rapid growth, and the capacities of existing outlets are unable to keep up with the demand. Without long-term planning capabilities, some enterprises only address problems when they become particularly prominent, leading to insufficient overall consideration when attempting to solve the immediate crisis, resulting in network planning and site selection that primarily serves local interests and hinders resource integration. Moreover, some express delivery companies employ a "three-year plan" or "five-year plan" system when constructing their networks, but given the rapid development of the express delivery business, there is a lack of real-time monitoring, predictive analysis, and dynamic planning capabilities, which is unable to fulfill user requirements and the increasing needs of the business.

9.2.5.4 Protruding Postal Monopoly Issue

The monopolization of postal franchises by public postal operators authorized by the country has become an urgent issue. Blind expansion of postal franchises will lead to two problems: first, it will have a certain impact on the improvement of efficiency of the entire express delivery companies in the country, and second, it will damage the legitimate rights and interests of other express delivery companies, thus impeding the formation of a market order with fair competition. The postal department has extensively employed its administrative power in the commercial field to establish market barriers, exclude competitors, reject the competition mechanism, and expand the scope of monopoly. Due to common interests, it is difficult to effectively supervise postal enterprises in the system. Furthermore, regulation of non-postal businesses is complicated due to conflicts of interests.

9.3 Analysis of Express Delivery Network Management Innovation Environment

9.3.1 Analysis of the National Environment

9.3.1.1 Advocacy for Innovation in National Policies

In recent years, China's express delivery industry has experienced rapid development. In 2017, the delivery volume reached 40 billion pieces, ranking first in the world for four consecutive years (In 2014, the delivery volume in China was 14 billion pieces, surpassing the United States for the first time to become the largest globally). In 2017, the General Office of the State Council issued documents, including "Opinions on Further Promoting Logistics Cost Reduction and Efficiency to Promote the Development of the Real Economy" (hereinafter referred to as "the Opinions") and "the New Generation Artificial Intelligence Development Plan", which proposed the development of smart logistics and the promotion of industrial automation upgrades. The combination of express delivery and intelligence has promoted further development of the express delivery industry, such as the establishment of intelligent distribution warehouses, production, and use of drones and unmanned vehicles. Additionally, the "Interim Regulations on Express Delivery" issued in March 2018 stated that the state encourages express delivery companies to use advanced technologies, such as automatic sorting equipment, mechanized loading and unloading equipment, express electronic waybill, and express information management system.

"The Opinions" proposed six policy measures. To begin with, institutional innovation should be strengthened, and the policy and regulatory environment should be optimized for coordinated development. Express business license procedures should be simplified, industry support policies should be innovated, and the inter-enterprise data sharing system and collaborative co-governance management mode should be improved. Additionally, planning and guidance should be reinforced, and the e-commerce express logistics infrastructure and express logistics network layout should be improved. An express logistics service system that accommodates the development of e-commerce should be constructed, and land should be provided for infrastructure construction. Furthermore, e-commerce distribution management should be optimized, and the operation of urban distribution vehicles in various places should be standardized to facilitate express delivery vehicles. Further still, the ability of basic level express delivery services should be enhanced, and smart express delivery facilities, express terminal services should be promoted, and service resources should be effectively organized and comprehensively utilized. Moreover, the construction of intelligent standardization should be strengthened, and modern information technologies and equipment, such as big data, cloud computing, and robots, should be applied in e-commerce and express logistics. The express logistics standard system should be reinforced, information sharing should be encouraged, resource allocation should be optimized (Zhang, 2018).

The application of innovative technologies for express delivery network management is an important factor in promoting the innovation of management capabilities of express delivery networks in the e-commerce environment. The continuous introduction of innovative technologies by express enterprises can promote overall development, and the continuous improvement of service quality and efficiency will drive the economic benefits of the e-commerce industry.

The nation has a fundamental role in the evolution of the express delivery network. The development of such a network cannot just be achieved by large express enterprises; rather, the nation should support it with policies to promote technology and system advances, in addition to helping the transformation and development of express enterprises. Therefore, relevant enterprises should also increase their capital investment, introduce new technologies such as big data and ensure efficient operation of the express delivery networks by utilizing the support of new technologies.

The express delivery industry should enhance their information management. Informatization is an inevitable requirement for development, and it can assist the express delivery network management in being more timely, accurately obtain various information resources, strengthen information transmission capabilities, and increase the operating cost of express delivery network management.

9.3.1.2 Exploration of Management Innovation Methods

Management innovation, a branch of management innovation, has received extensive attention in academic circles in recent years. Van de Ven (1986) was the first to propose that innovations are new ideas that arise and develop over time through interactions with others within an institutional order. This definition focuses on four fundamental factors (new ideas, personnel, interactions, and institutional context). Buchmiller & Connell (2003) claimed that a successful enterprise should extend the entire innovation activity to all aspects, from the initial innovation demand to innovation exploration, the stimulation of innovative ideas, innovation cooperation and investment, and innovation reporting and communication. Halman & Keizer (1994) proposed a new approach to innovative projects, which is an improvement on existing risk methods used in product innovation projects, such as potential problem analysis, failure mode, and impact analysis. Zhang & Jia (2017) classified users' needs and preferences for the first time, and proposed a product innovation method for importance evaluation based on that.

Research into management innovation in China is increasing; however, there are gaps in the research into management innovation methods. Zhang et al. (2014), Xu (2016), Li et al. (2017) and Zhao et al. (2017) have all explored the innovation behavior and methods of enterprises, yet none of them proposed a method with practical benefits. Jiang & Li (2015) argued that, in the management innovation of the express delivery industry, changing the competition mode and improving technological innovation would be necessary for the transformation and upgrading of the national express delivery industry. Wu (2017) emphasized the need for the express delivery industry to implement collaborative innovation and development. Further exploration of the specific management innovation methods of the national express delivery network is required.

Most innovations in China's express management are confined to national policy and operational models with the sole purpose of generating profit through sales. These innovative activities are often highly replicable and imitate one another. However, there is a serious lack of production organizational structures, collaborative innovations, and adaptive management methods that are able to truly reflect the core competitiveness of the enterprises and are not easily imitated or replaced. This is something that express enterprises must strive to explore.

Many successful enterprises around the world focus on learning new methods and regularly organize management training. Informal discussions and learning among the same industry occur periodically, and they develop a set of the most efficient methods in the application of innovative methods. Enterprises then improve and refine the innovation method according to their own development status and form an innovation system of enterprise-specific management modes. This approach, to some extent, promotes the development of theoretical innovation, implementation methods, and research tools for management modes.

When an innovative method is successfully applied to an enterprise, it is not only the method itself that is important, but also the staff who have the capacity to maximize its effectiveness. This is an essential element for enterprises to successfully complete the innovation process. Countries across the world have always placed great emphasis on the cultivation of talents, and enterprises invest heavily in the introduction of professionals and the provision of systematic training. With the continuous accumulation of experience in the application of innovative methods, some of them will be systematized and effectively integrated into the system, creating a unique training system and tools for the enterprise.

9.3.2 Analysis of the Industry Situation

For over two decades, the combination of e-commerce and the Internet has clearly been an essential feature and long-term trend of economic and social development. However, at present, there is an absence of standardization in express management in our country, resulting in a chaotic express market, with many issues remaining unresolved.

9.3.2.1 Irregular Service Management

In December 2018, the National Post Bureau and provincial (regional and municipal) postal administrations department accepted a total of 220,110 consumer complaints through the "12,305" postal industry consumer complaint hotline and complaint website. Of these, 214,040 were related to express services, accounting for 97.2%

of the total number of complaints. Common issues behind these complaints include poor customer service attitude, delayed delivery, loss of express items, and damage to express items. The current express enterprises in the country are characterized by irregular customer service standards, low service quality of deliveryman, ambiguous corporate culture, low prices, and vicious competition. Moreover, there are many express enterprises operating without legal business qualifications, which further contributes to the market pricing mechanism's destruction and the recruitment of employees with poor service awareness, in violation of express service management principles.

9.3.2.2 A Single Employee Training Mode

The express delivery industry does not necessitate the educational qualities of its employees. When recruiting employees, as long as the candidates are able to fulfil the tasks of express distribution and delivery through the use of transport tools, they fulfill the recruitment criteria. New employees are only given a basic introduction to the process, with no comprehensive service standards or post-specific training. When the express enterprise is faced with a major e-commerce event and the number of express deliveries suddenly increases, their requirements are even lower. In general, when temporary recruiters are handling e-commerce activities, new recruits do not even receive a basic introduction to the process, instead being expected to do their job directly, which consequently leads to poor service quality and low service efficiency in the express delivery industry (Wang, 2017). Therefore, it is widely accepted that the job is simple and requires hard physical work, and that express delivery employees do not require a high level of education.

Relevant personnel of express delivery network Management should be proficient in the theoretical knowledge and business operation skills of the entire department and must possess a good understanding of the entire express delivery network System. Additionally, managers must coordinate the arrangement of relevant personnel and vehicles at Express Call Centers, Express Pick-up and Delivery Outlets, as well as transit stations and transportation routes at all levels. Express Enterprises must also have relevant knowledge of economic trade, Express Management, and mature Logistics Control Capabilities. The comprehensive quality of talents is an important factor in determining an enterprise's ability to obtain core competitiveness. Each year, a large number of university graduates enter the workforce with highly educated backgrounds. However, they fail to meet the specific requirements of Express Enterprises, which includes a combination of theory and practice, related operations, as well as experience and knowledge reserves. It appears that, although new employees are recruited from universities, they are not sufficiently qualified to work independently.

9.3.2.3 Lack of Exclusive Transportation Equipment

In the implementation of distribution services, express enterprises mostly utilise electric tricycles and minivans, which provide high flexibility, as their main means of transportation, with electric tricycles being the most common. Nonetheless, the safety of the transportation equipment cannot be disregarded. When the express storehouse is overwhelmed, the tricycle will carry fewer parcels, and when the round-trip distance is long, the delivery efficiency will be impacted, thereby increasing the probability of express delay. Having exclusive transportation equipment for express enterprises can not only improve transportation efficiency effectively, but also ensure the personal safety of deliverymen and decrease the damage rate of express deliveries. Nevertheless, the vehicle manufacturing industry has not been able to create delivery vehicles that meet the requirements of the enterprises for transport vehicles. Moreover, the pre-existing transportation information systems of the enterprises have issues such as outdated system versions, incompatible system functions, system program crashes, and daily business management freezes. These difficulties will result in chaotic system data, an inability to manage and control vehicles precisely and efficiently. All of these lead to a low distribution efficiency.

9.3.2.4 Unbalanced Development of the Express Delivery Industry

In Chapter 3, we systematically analyzed the geographical distribution of express enterprises in China and concluded that the level of economic development in different regions of the country is varied. There is an obvious regional disparity in the demand for express services which causes an uneven distribution of express service outlets in various regions. The areas with relatively concentrated express business volume are the eastern coastal areas, where express turnover is relatively large and domestic express enterprises have established extensive service outlets in major and medium-sized eastern cities. On the other hand, economically underdeveloped areas mostly lack express outlets in small and medium-sized rural towns, and even the established e-commerce service sites are unmanaged.This imbalance has caused corresponding regional differences in the service quality of express enterprises, thereby restricting the development of e-commerce in backward regions as well as having a negative impact on the coordinated development of the express delivery industry across all regions.

9.3.2.5 Necessary Improvement of Infrastructure Construction

On January 23, 2018, the General Office of the State Council issued the "Opinions on Promoting the Coordinated Development of E-commerce and Express Logistics", which mentioned the construction of express infrastructure, including related warehousing, distribution, delivery, as well as terminal express smart lockers and comprehensive service places. This indicates that the state has placed great importance on improving express logistics infrastructure. Generally, the current state of the logistics industry in our country is lagging behind developed countries, particularly in terms of logistics infrastructure. Despite the express delivery industry's rapid growth throughout the country, there are still significant deficiencies in transportation, communication, management, handling and warehousing. Currently, with the rapid development of e-commerce, the express delivery network infrastructure is unable to meet the development requirements. Therefore, it is essential to construct a complete set of express delivery network system in order to achieve management innovation for express delivery network.

9.3.2.6 High Cost of Technology Application

The cost of technology application is mainly reflected in two aspects. Firstly, due to the imperfect management mode of the national express delivery industry, businesses still rely on traditional management methods for innovation, resulting in increased costs for human resource management and decreased management efficiency, especially with the expansion of the express delivery industry. The other is that the existing domestic express management system is not able to keep up with the technology of the international forefront. When applying technology, the overall coordination of the logistics system must be taken into account. Some express enterprises have adopted advanced technology for management innovation, but the relevant technical personnel and supporting equipment have not been able to match it, thus impeding the relevant functional advantages of technical equipment and increasing operating costs for express enterprises significantly.

9.4 Summary

Clarifying the classification of the current management mode of express delivery networks and exploring the shortcomings of the current express delivery network management is the premise and foundation for the reform and innovation of express delivery network management. In this chapter, the current three management modes of express enterprises are analyzed: direct-operated mode, franchise or agency mode, and self-operated mode. It can be observed that the business modes of express enterprises in our country are predominantly direct-operated mode and franchise mode. The direct-operated mode is beneficial for management; it enables enterprises to ensure the quality and standard of express services, and facilitates financing. The franchise mode is more focused on expanding and maintaining operations through the involvement of participants, who are bound together by a franchise agreement, yet remain independent when it comes to economic management. Following this, the relevant factors that currently promote the reform of the express delivery network management mode are further analyzed and summarized, mainly including market factors, price factors, logistics facilities, distribution location and related policy factors. Finally, the necessity of express delivery network management innovation is discussed in detail from the perspective of the national environment and industry situation. Generally, this chapter provides the basis for subsequent reform and innovation of express delivery network management.

References

Buchmiller, J. L., & Connell, C. L. (2003). Innovation engine portal method and system, US, US 20030187706 A1.

Gao, B., & Tao, B. G. (2013). *Introduction to express service*. Beijing: Posts & Telecom Press.

Halman, J. I. M., & Keizer, J. A. (1994). Diagnosing risks in product- innovation projects. *International Journal of Project Management, 12*(2), 75–80.

Jiang, B., & Li, J. (2015). Domestic market size and industrial upgrading of China's express industry. *China Business and Market, 29*(1), 77–83.

Jiang, L. L. (2018). *A spatial analysis of the express delivery network of "China Railway Express"* [Master dissertation, Shanghai Normal University].

Li, R. D. (2018). A study on the business model of rural express delivery enterprise. *Agricultural Science-Technology and Information*, (14), 68–71.

Li, R., Liu, L., & Zheng, X. N. (2017). Supplier-facilitated inventory transshipments: An innovative approach based on firm practice. *Journal of Management Science, 30*(6), 92–103.

Lin, S. S. (2016). Research on express industry in the e-commerce environment—in the case of SF. *Logistics Engineering and Management, 38*(8), 81–82.

Tian, X. Y. (2017). Analysis of advantages and disadvantages of Jingdong logistics distribution mode and its countermeasures. *Modernization of Management, 37*(6), 92–96.

Van de Ven, A. H. (1986). Central problems in the management of innovation. *Management Science, 32*(5), 590–607.

Wang, F. M. (2017). On the current problems and opportunities in the development of China's express delivery industry. *Ability and Wisdom*, (16), 265.

Wu, P. (2017). Research on the development path of express delivery industry from the perspective of collaborative innovation. *Jinan Journal (Philosophy & Social Sciences), 39*(03), 25–29.

Xu, S. Q. (2016). The fuzzy comprehensive evaluation model and empirical analysis of the effect of enterprise application innovation methods. *Science and Technology Management Research, 36*(6), 79–84.

Yu, Z. H., & Zhang, C. (2014). A comparative study on the business models of express delivery in China. *China Market*, (49), 16–17.

Zhang, B. Y. (2018). The General Office of the State Council issues *the opinions on promoting the integrated development of E-commerce, delivery and logistics. Computer & Network, 44*(5), 09.

Zhang, F. L., & Jia, C. X., (2017). Products innovation method based on classification and importance evaluation of user needs. *Packaging Engineering, 38*(16), 87–92.

Zhang, L., Qi, E. S., & Chang, Q. (2014). Evaluation of management innovation methods types for manufacturing enterprises: An empirical analysis based on support vector machine. *Studies in Science of Science, 32*(11), 1747–1753+1760.

Zhao, G., Liu, B., & Li, Z. B. (2017). The influence of entrepreneurial behavior of incubation enterprises on entrepreneurial performance and promoting policies—A case study of Tianjin. *Enterprise Economy, 36*(1), 88–95.

Chapter 10
Reform and Innovation of Express Delivery Network Management

10.1 Reform of the Express Delivery Network Management Mode

China's express delivery network has various issues in the management process at different levels, making it highly significant to alter the traditional management mode. By analyzing the express delivery network model and its changes of relevant influencing factors, the authors proposed an operational model which combines direct operation and franchised models.

The purpose of reforming express delivery network management is to improve management capabilities, increase operational efficiency, reduce operating costs, and enhance customer satisfaction by altering management principles, network systems, express transportation, and customer service. The success of this reform can be assessed by examining the performance of express delivery operations costs, the speed of transformation, and whether customer satisfaction has been improved. However, reforming express delivery network management is a difficult endeavour. From a broad perspective, the express delivery enterprise organization involves a wide range of functional departments, encompassing all aspects of the enterprise, including transportation, warehousing, packaging, distribution, sales and finance. From a narrow perspective, the express delivery network comprises numerous express delivery call centers, express delivery pickup and outlets, transit stations at all levels, and the entire delivery transportation route. The reform of express delivery network management involves all aspects in order to improve the operational performance of local express delivery network management and the management of the entire enterprise organization. Therefore, the transformation of the express delivery network management mode should be conducted from the perspectives of management mechanism and industry technology. The overall block diagram is illustrated in Fig. 10.1.

P. Li and J. Wu, *Research on Innovation of Express Delivery Network Management*,
https://doi.org/10.1007/978-981-99-8982-9_10

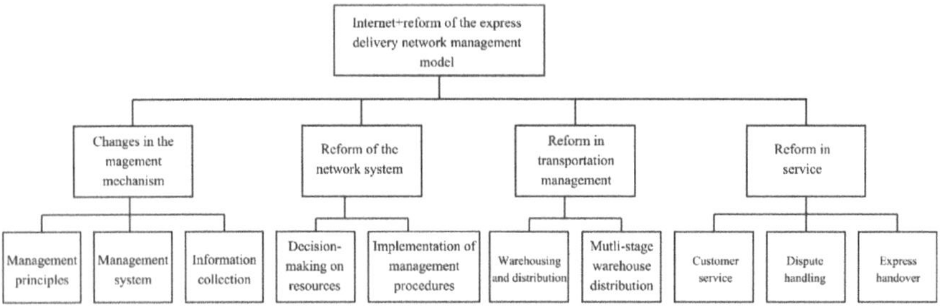

Fig. 10.1 The overall block diagram of management mode reform. *Source* The Author

10.1.1 Changes in the Management Mechanism

10.1.1.1 Updating Management Principles

In the digital economy era, the flourishing of e-commerce has changed the business environment, management methods, and production operations of traditional express delivery enterprises, thus forcing corresponding changes in their management objectives and concepts. To stay competitive in the express delivery market and meet customer demand, these enterprises must quickly adapt to the continuous reform and upgrading of the current industrial structure, ensuring the proportion of service-oriented, technology-oriented, and innovative enterprises increases through management transformation. Additionally, they must pursue independent adaptability, innovation, and maximum customer satisfaction as management concepts, reflecting the ideals of the e-commerce era. The core of the overall management thinking of express delivery enterprises should be to demonstrate express delivery and customer-centricity. The key to the reforms of the express delivery network is to prioritize express delivery and customers, and to pay attention to services and customers in all stages of the express delivery network process. Management should not only traditionally emphasize process, system, and results, but should likewise be more customer-oriented, focusing on employees and emphasizing corporate culture.

10.1.1.2 Improvement of the Management System

The marketization of express delivery service prices in China is low, and problems such as vicious price competition, weak price supervision, and pricing mechanisms need to be addressed. From the perspective of "Internet Plus" and modern management concepts, the relevant regulations of the express delivery industry should be improved. First, taking into consideration the current situation and trend of the express delivery industry and the Internet, a corresponding institutional system should be established to ensure successful implementation of the management mode in all aspects (Feng et al., 2019). Second, the modern management mode is constantly evolving, thus necessitating the continuous revision and enhancement of the ability to implement, conduct, and self-renew.

10.1.2 Reform of the Network System

Express delivery enterprises should establish a data center in order to collect and analyze data, which is the foundation of the express delivery network management. This is the starting point for corporate decision-making and operation. To achieve this, the enterprises should build a standardized data management mechanism and conduct systematic training for staff related to enterprise and technology management. Furthermore, an open management mechanism for information sharing within the enterprises and among various departments should be established in order to facilitate the cross-departmental and cross-functional flow of data and information results. Through this, the management decision-making of express enterprises can become more intelligent and unified.

Express delivery network management should incorporate external resources to facilitate the introduction of Cloud service technology into their management system. Domestic software enterprises have developed Cloud service basic platforms for many express delivery enterprises, such as Jingdong. These platforms can effectively contribute to the overall express delivery network management. All enterprises should set realistic development goals, optimize application processes, and apply data to all operational links. This can substantially reduce the consumption of human and material resources. In order to fully capitalize on the advantages of new technologies such as big data and cloud computing, express delivery enterprises should integrate their operations into a network to form a collective advantage. By doing so, big data technology can be applied to all areas of the express delivery network, including data analysis, goods preparation, warehouse storage, and transportation management. This can be integrated into a unified network platform, allowing the relevant links and data of each area to be shared, thus enabling innovation in express delivery network management and improving the efficiency of express delivery business operations (Li, 2014).

In this reform of the express delivery network management mode, with "Internet Plus" as the main line and leveraging the big data platform, technological changes will be implemented in order to reach the desired effect. The development of information technology is directly linked to the transformation of the management mode of the express delivery network. Information technology has been effectively applied to express delivery network management, such as identification and collection technology, as well as transmission technology. It is essential in all aspects of the express delivery network management mode reforms. Consequently, technological change should be conducted from three perspectives: information collection, resource decision-making and application execution.

10.1.2.1 Information Collection

Information collection involves perceiving, collecting, transmitting, storing, and processing information. To facilitate information collection, objects must be collected. The express delivery industry can construct a "Cloud warehousing" service based on socialized logistics. Different logistics businesses have separate locations in different cities, providing physical warehouses suitable for various commodity storage. Each physical warehouse is subdivided into multiple logical sub-warehouses. Additionally, it provides quick and dependable "nearest warehousing and local distribution" services for different e-commerce companies. This solves the problem of cross-regional industries' network storage and facilitates the collection and organization of different information data.

10.1.2.2 Decision-Making on Resources

The decision-making module optimizes the logistics information obtained from the information collection module through intelligent technology, provides corresponding decision support through analysis and prediction, and effectively integrates information resources with the help of modelling and simulation technology to achieve macro-control of the entire logistics process and coordination between tasks. By leveraging rules and algorithms, each part is allocated the optimal path, which can push massive and varied tasks from any platform source to the nearest "Cloud Warehousing" warehouse at the present time, automatically and accurately, thus improving the efficiency of express network management.

10.1.2.3 Implementation of the Management Procedures

The Terminal Application Module primarily implements particular actions based on the received resource decision-making information in order to assist express enterprises in completing logistics operations efficiently. This part also possesses the function of terminal feedback; it receives and implements relevant instructions, and feeds back to the Resource Decision-making Module in a timely manner based on the application execution situation. Subsequently, the Resource Decision-making Module feeds back to the Information Collection Module, and the Information Collection Module makes suitable adjustments based on the feedback information. Thus, the entire Express Delivery Network Management Module can make optimal adjustments while operating continuously.

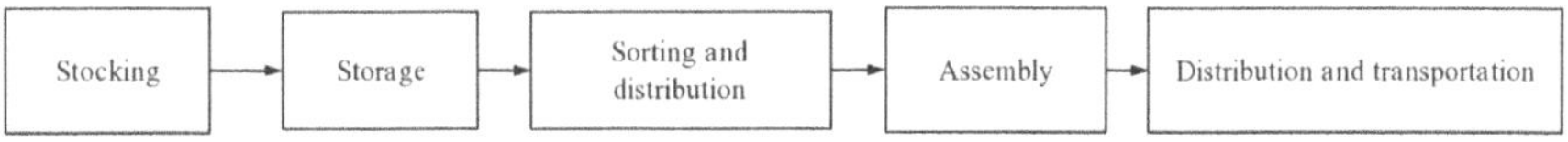

Fig. 10.2 Express delivery distribution process

10.1.3 Reform in Transportation Management

10.1.3.1 Warehousing and Distribution Optimization

There are five stages in the entire express delivery distribution process. In order to improve management during transportation, all components, from stocking to delivery and transportation, must be taken into consideration. Specific measures to be taken include the addition of professional staff to check the express delivery upon scanning. Their primary task would be to carefully inspect the express delivery package for damage once it has been transported to the sorting center for scanning. In the case of damage, the package information should be immediately fed back to the enterprise, thus reducing the likelihood of the deliverers and the enterprise being held responsible for the damage, and enhancing customer service satisfaction for express delivery enterprises.

In the part of warehousing and distribution optimization, mainly through the information platform of the electronic network, the express delivery service demand information between consumers and enterprises is collected (Qi, 2019). And the information obtained is used to calculate and analyze the optimal distribution solution for the express delivery chain, ensuring a more convenient and efficient logistics mode, improving transport efficiency and reducing logistics costs. In this distribution chain, traditional storage technologies and equipment can be replaced by intelligent devices. Smart devices can be selected: shelf shuttle equipment, automatic sorting robots, unmanned forklifts, wearable devices (hands-free scanning devices, AR smart glasses), tasked with sorting and handling express delivery, shelving, etc. In this way, the distribution chain realizes the intellectualization of the whole warehousing process, effectively reduces the human resource investment in the warehousing and distribution part, improves the efficiency of logistics distribution, and establishes and improves the intelligent warehouse management system. The distribution process of express delivery is shown in Fig. 10.2.

10.1.3.2 Multi-Stage Warehouse Distribution System

According to the transportation route and logistics cost, the multi-stage warehousing system determines the optimal warehousing outlet to improve the distribution rate of goods. This system integrates the storage, sorting and scanning of logistics information, and cargo distribution, including warehousing centers, outlets, and network systems. The system can automatically generate different distribution schemes and exclusive logistics distribution systems based on the attributes of the distributed commodities, transportation equipment, and human resources. The high efficiency of the distribution system is attributed to the adoption of the co-flow and intensive processing of express delivery orders, as well as the fast response to customer orders. For instance, the system divides the distribution levels based on the customer's needs, the freshness of the goods to be delivered, and the volume of the express, and calculates the optimal transport path, thus enhancing the circulation efficiency of goods and decreasing the logistics and distribution costs of orders. Maintaining the efficiency of the system requires accurate inventory allocation, which can be achieved by building an inventory management system for sharing and analyzing historical data in different regions through information sharing in order to predict the necessary inventory for each area. By integrating independent information into relevant logistics storage resources, the loading rate and distribution time of express delivery vehicles can be improved (Zhang & Zheng, 2019).

10.1.4 Reform in Service

Service-oriented management reform is a key step in the successful transformation of an enterprise. It requires that customers' opinions and suggestions be regarded as the most important factors, and the enterprise should be managed continuously based on that. On this basis, the innovation of express delivery network systems and production and operation management should be promoted. The service is the communication behavior between express delivery enterprises and customers in direct contact. These enterprises must abandon the traditional management principles and guiding ideologies, no longer be limited to short-term interest needs, take customer-focused commitments and focus on improving user satisfaction. Through express delivery, customer service and dispute resolution, the service quality can be improved, and the brand image can be established to enhance user loyalty.

Express Delivery

Express delivery is a fundamental component of the handover process. This process involves the direct contact between couriers and customers, which is still the primary express method today despite the emergence of express delivery pick-up cabinets, Cainiao Stations and other pick-up and delivery centers. Express handover consists of three parts: express delivery, distribution and signature, and serves as a stage for express delivery enterprises to interact with customers. The performance of couriers is an important factor in determining how customers view a particular express delivery enterprise.

Customer Service

Customer service is the focus of service transformation, and express delivery enterprises should strive to broaden the channels for providing help to customers. In the age of the Internet, telephone consultations are no longer sufficient to meet customers' needs. Therefore, the enterprises need to improve further new service models like online consultation, online customer service and customer message, etc. and carry out professional and systematic training in order to reduce the prevalence of unanswered consultations, incomplete reply information, and inadequate customer service attitudes.

Dispute Handling

Dispute Handling is a key point. It is essential to improve user satisfaction and establish a positive corporate image for enterprises. Consequently, the express delivery enterprises should pay close attention to the recently released national laws and regulations, and abide by the national standards when conducting business operations. Additionally, they should strive to reduce disputes resulting from violations of regulations and disciplines, and establish an effective dispute-resolution mechanism to ensure that disputes between parties can be quickly and correctly resolved (Qin, 2016).

10.2 Innovative Architecture Model for Express Delivery Network Management

The emergence of e-commerce has greatly stimulated the development of the express delivery industry, which has had a significant impact on traditional Chinese economic development. Some companies have paid a hefty price for transformation and innovation. Express delivery enterprises mainly concentrate on internal management modes and external business environments for management innovation.

Express delivery network management is confronted with threats from the rapid evolution of transportation equipment and network technology, the continual growth of operational risks and industry competition, the rising competitive pressure and the diminishing customers loyalty. Under the effect of e-commerce, express delivery enterprises must alter their traditional business philosophy, management approaches and ideology, foster innovative thinking and concepts,, and enhance innovative management.

Tian (2016) used the growth and rise of Huawei as a case study to explore the innovation of enterprises. He suggested that Huawei's innovation is comprehensive, with the most significant form being concept innovation. Zeng (2013) argued that the innovation of management concepts can have a deep and lasting impact on all innovation activities. This implies that managers should accurately identify new changes in the world, comprehend the rules of global development and potential trends, and create new ideas and concepts in a meticulous manner.

For express delivery enterprises, the impetus and basis for continuous innovation lies in the innovation of management concepts, which should be adjusted to correspond to the development of society and encourage new ideas and management concepts. Subsequently, based on extensive research and in-depth analysis, and

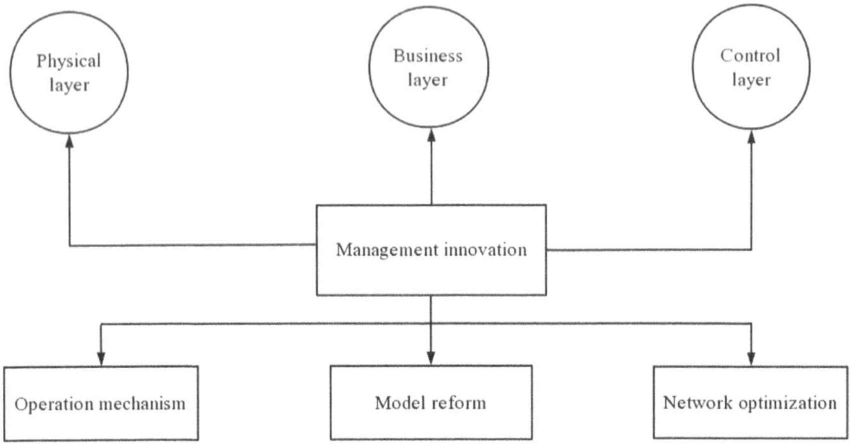

Fig. 10.3 Express delivery network management innovation model. *Source* The Author

guided by the innovation theory of enterprises, the author has established an innovative network management mode for express delivery, which takes into account network optimization, operational mechanisms, and mode changes. This is depicted in Fig. 10.3.

The progression of management innovation activities is inextricably linked to an efficient and comprehensive management system. Through the implementation and operation of such a system, express delivery enterprises can cultivate the development of management innovation and create a proprietary enterprise management approach to gain a competitive edge against other businesses.

Management innovation is essential for the development of enterprises and the generation of economic benefits. Only through management innovation can enterprises reach their market value and achieve economic benefits.

Enterprise innovation is composed of two elements: "technical innovation", which includes new technologies, ideas, materials, products, and other "hardware" elements; and "management innovation", which encompasses new organizations, markets, business management modes, production methods, processes, and services (Jing & Niu, 2014). Technological innovation and management innovation form the core of an enterprise innovation management system. By innovating both "software" and "hardware," a "multi-management innovation ecology system" is created, as depicted in Fig. 10.4.

Scholars across the world have conducted research on enterprise management systems from multiple angles. Wang (2015) and Chen (2017) examined the current financial management situation in the express delivery industry and proposed improvements and optimizations. Nevertheless, there is still a need for further detailed research on the operational modes and methods of express delivery network management innovation in academia.

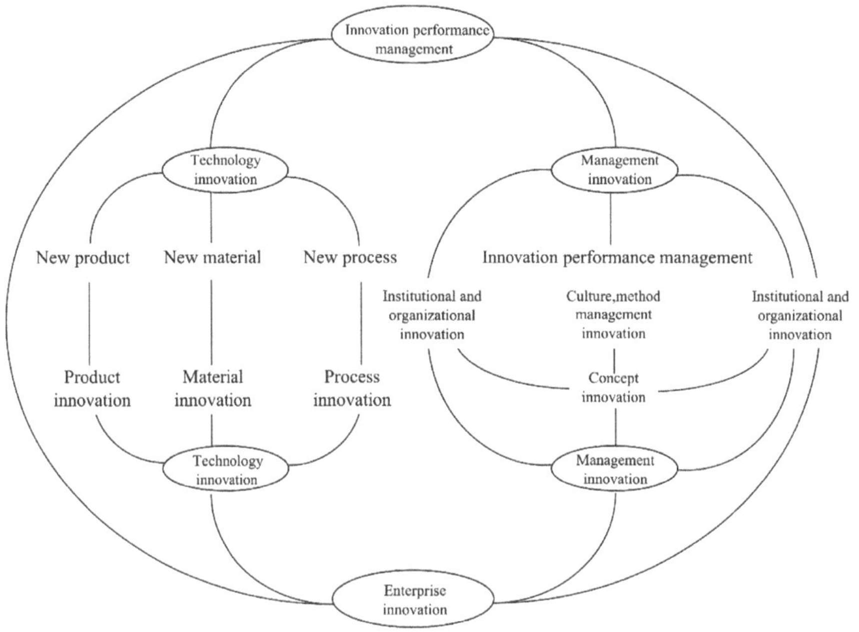

Fig. 10.4 Multi-management innovation ecology system. *Source* The Author

10.3 Innovative Express Delivery Network Management Model

In the preceding chapter, the existing express delivery network management modes were summarized and classified, and the strengths and weaknesses of each management mode were discussed in depth. In combination with the reform of the express delivery network management mode in the first and second sections of this chapter, as well as the innovative ecological architecture mode, the author constructed a multi-ecological express delivery network management mode with adaptive qualities.

10.3.1 Multi-Ecological Express Delivery Networks Management Model

10.3.1.1 The Current Elementary Model

The current express delivery management mode in China can be classified into the direct-operated, franchise and self-operated models. Under the direct-operated model, the headquarters centrally manages ownership and operational rights, exercising centralized leadership and unified management. All direct-sale stores are required to implement standardized operations and management. This model is advantageous due to its ease of use. The headquarters develops new technologies and projects in a unified fashion, applying them to the overall business to achieve an optimal allocation of shared resources. Sub-enterprises are able to share the capital chain, information and management experience held by the headquarters, while the cooperation and support among enterprise members help the headquarters to obtain the most effective market information. The direct-operated model is characterized by the headquarters having direct control over each sub-enterprise (Xu, 2016); however, this control has certain drawbacks. On the one hand, it has generated strong capital for the headquarters, yet it requires substantial investments in transportation, outlets, warehousing, and personnel, and takes a considerable amount of time to build a corporate brand, which may lead to a capital chain breakage crisis. On the other hand, it also reduces the sovereignty of individual members of the direct-operated model enterprise, resulting in diminished enthusiasm, creativity, initiative, and efficiency.

In Chapter 9, we have conducted a thorough analysis of the franchise model. Under this model, the headquarters deploys regional general franchisees to appoint subcontract franchisees, with the most basic level being regional express delivery outlets. Express delivery enterprises operating under the franchise model are not subject to interference from the headquarters, allowing them to independently recruit employees, set prices for express products, and devise marketing strategies. Therefore, the franchise model facilitates rapid development of the company's agency network, enabling it to cover the entire country in a relatively short period of time and quickly build the company's brand. The franchise model has the advantage of relatively low management costs, since corporate headquarters and franchisees are independent legal persons and are only responsible for their respective enterprises. However, this method weakens the horizontal control and management of each franchisee by the headquarters and makes it difficult for sub-franchisees to adapt to the requirements of integrated management among express delivery outlets (Yu & Zhang, 2014). In the long-term, this kind of management mode is prone to the manifestation of independent outlets, loose outlet control, weak execution, and even a low-price monopoly market and vicious competition, which is not conducive to the development of the express delivery industry.

10.3.1.2 Multi-Ecological Express Delivery Network Management Model

The multi-ecological management model combines the benefits of direct operation and franchise models. It is formed by applying modern technologies such as artificial intelligence, Internet of Things, and big data. This model encompasses direct management led by headquarters and unified command, and allows for full utilization of advantages in talent training, technology research and development, and information management. Moreover, the franchise model's advantages are included, enabling employees to work for a certain number of years or a certain position, thereby increasing employee enthusiasm and easing the company's capital turnover to a certain extent. The main innovations of this multi-ecological management model are: Internet of Things + big data platform construction management model, network data mining + information resource optimization management model, and network system framework + customer core personalized service management model.

1. Interent of Things + Big Data Platform Construction management model

The Internet of Things (IoT) is an information exchange network that interconnects every element of our lives, from people to people, people to things, and things to things. It is utilized in enterprise resource management to create an intelligent enterprise information management platform, which consists of four distinct parts: the infrastructure and perception unit, the network unit, the data information processing unit, and the application unit. The infrastructure layer includes all the data information elements of the platform, while the perception layer identifies, collects, and monitors various data information in the infrastructure layer through chips, wearable devices, sensors, and cameras. This layer also enables activities such as the scanning and uploading of real-time logistics information, and the monitoring of any contraband in an express shipment. The network unit mainly comprises network equipment, networks carried by the Internet of Things, and mobile communication networks, among others. The main data sources are obtained from the perception layer and access to external electronic platforms. Accessing external data increases the amount of data measured on the platform, improving the accuracy of the unit measurement and the precision of the obtained information, as well as enabling the efficient sharing and utilization of information resources between platforms (Liu, 2013). Infrastructure and perception units, network units, and data information processing units coordinate with each other, and the results are presented to the customer via the application layer. The application layer also provides valuable decision-making information to enterprises, such as auxiliary decision-making and decision-making suggestions, which enhances the intelligent informatization degree of enterprise resource management under the "Internet of Things + Big Data" model and achieves the dynamic balance of the information ecosystem.

2. Network Data Mining + Information Resource Optimization management model

In the process of innovating express delivery network management, global data informatization presents both an opportunity and a challenge compared to building a new model. The primary aim of the data informatization era is to extract valuable information from data, and data mining and data analysis are the key steps to fully utilize the value of data. The Network Data Mining + Information Resource Optimization management model uses data mining, data intelligent analysis, and other technologies to identify the intrinsic value of information resources, filter and integrate effective data, and present it to users; this not only enhances the user experience and optimizes the quality of information services, but also provides express delivery companies with better quality. This model enables efficient use of data to realize the maximum potential value of data mining and to continuously optimize it.

3. Network System Framework + Customer Core Personalized Service management model

Personalized customer service is essential for the improvement of customer satisfaction and the establishment of a positive corporate image. User behavior data holds a great deal of potential information. By analyzing customer history data, the behavior characteristics of users can be determined and demarcated, allowing for an accurate understanding of the customer's needs and goals. The network system framework and customer core personalized service management model utilize data mining and data analysis technology to establish a Web mining analysis mode. This mode consists of five modules: web page click behavior module; page jump behavior module; online service demand module; user demand satisfaction module; behavior correlation module. The modules in the model optimize the service interface through various user behaviors and improve the service content according to individual user needs. The network system framework and customer core personalized service management model identify key nodes in information dissemination through established methods; this enables the excavation and identification of the real needs of users, efficient and accurate pushing of service information (Yu et al., 2016), increased impetus and influence of Internet information, and user-centered personalized service management in data information service (Chang, 2013).

10.3.2 Adaptability of the Plural Ecological Management Model

The purpose of the constructed plural ecological management model is to create a connection between enterprises, employees and customers by utilizing new technologies such as data mining, intelligent analysis and web mode. Consequently, enterprises can formulate rational and perceptual strategies, construct an enterprise ecological model framework, and devise corresponding enterprise management strategies, thus enhancing their competitiveness (Wang et al., 2017). The attempt to construct a pluralistic ecological management model is based on ecological thinking, which internalizes the express delivery enterprise into a multi-ecosystem comprising the management of the evolutionary process of identification, planning, implementation, evaluation and self-renewal of the express delivery enterprise system (Wang, 2011). This model is a collective of express delivery companies and stakeholders, with two main types of relationships present: the vertical relationship and the horizontal relationship. The vertical relationship consists of express delivery companies and their suppliers (primarily e-commerce), consumers, and market intermediaries; the horizontal relationship pertains to the relationship between companies, competitors and stakeholders in terms of both competition and cooperation (Cao, 2006). The relationships between modules of the multi-ecosystem management model are depicted in Fig. 10.5.

The management model of the plural ecological express delivery network is characterized by its self-adaptability, which consists of two components: self-circulation and self-renewal. In constructing the management model of the platform of IoTs +

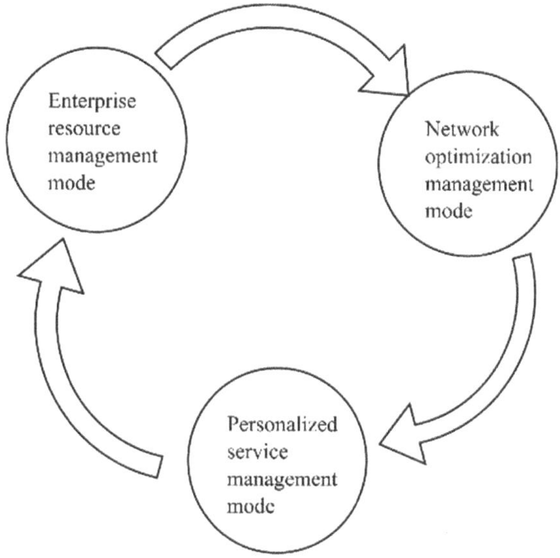

Fig. 10.5 Relationships between modules of multi-ecosystem management model. *Source* The Author

big data, the intelligence and automation of the management of information resources are improved through the use of IoTs and big data technology to achieve the self-circulation of the ecological management model and to realize the ecological balance of the information system. Through data mining technology, cloud computing, and intelligent analysis technology, the network data mining + information resource optimal management model is used to deeply extract data that is difficult to digitize and to uncover its hidden value. The express delivery network system maximizes the value of the mined data through established algorithms. Data mining technology transforms customer behavior data into "effective data" which are then applied to the express delivery network system, optimizing and improving it to form a data cycle that helps the management system of the express delivery network to achieve an ecological balance between its own data and system optimization. In the section of the management model of personalized service that takes the customer as the core and utilizes a technical framework, new technology is utilized to establish the distribution mode of time and space of customer's demand, the mode of column correlation, and a recommendation algorithm is employed to optimize the customer service interface for different users. Simultaneously, it is equipped with information service delivery technology based on the public opinion data of the entire network, which provides efficient and accurate service messages to users (Han & Wei, 2019). This not only increases the impetus and influence of internet information but also enhances the customer's sense of experience, resulting in user-centered personalized information services. The influence of the internet and the customer's sense of experience will assist express delivery enterprises in continuously improving their technical frameworks and modes, thus maintaining the ecological balance of the user's personalized service management model. Overall, the management models of enterprise resources, network optimization, and personalized service are mutually promoting. User behavior data is recycled among these modules, so when the customer behavior data is updated, some components of the enterprise resource management model are also updated, such as the network platform, network system optimization, and the platform providing personalized service to customers. As a result, the ecological management model has a self-updating feature.

10.4 Innovative Strategies for Express Delivery Network Management

10.4.1 Connecting the Logistics Networks

Connecting the logistics networks refers to combining the logistics of express delivery networks to form an extensive and comprehensive network. This is primarily achieved through the Small World Model, which enables customers to satisfy their multiple needs in an efficient and cost-effective manner. The proportion of effective express delivery logistics network information serves as the connection degree.

In conclusion, further analysis can be conducted in order to classify express delivery network management models and explore reform and innovation in express delivery network management. It is found that the continuity of the integration of the express delivery industry and the matching of the express delivery network platform make the connection of the express delivery network the core of the management coordination and optimization of the express delivery industry. Integration strategies of the express delivery network mode are provided from three perspectives, including related resources of the express delivery network, corresponding functions of express delivery network management, and goals and tasks of express delivery network management. The roles of the three in the express delivery network management model are interaction and feedback to each other, as illustrated in Fig. 10.6.

10.4.1.1 Resource Integration

From an industry-wide perspective, there is an ineffectiveness and lack of interactivity in resource docking among express delivery companies. To increase the scope of "network virtualization", it is necessary to discuss express delivery transportation and efficiency from the viewpoint of "resources" and to integrate resources within the express delivery industry. For the management of the express delivery network, the express delivery industry should consider resources strategically and comprehensively, including physical resources, labor resources, capital financial resources, and information resources. Then, they should carefully distinguish the resources that

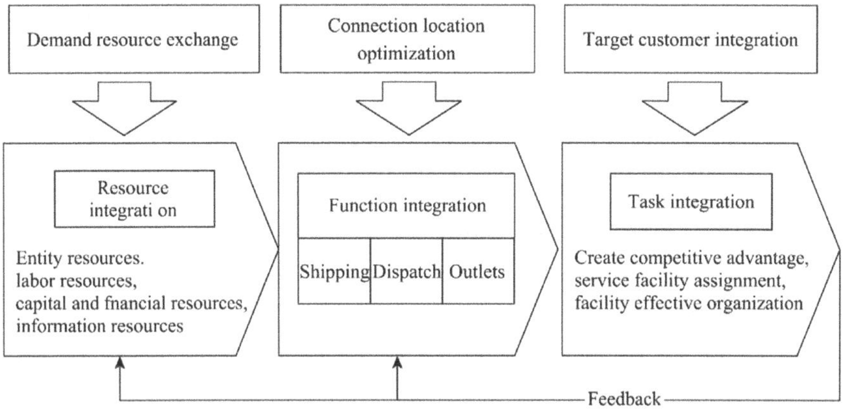

Fig. 10.6 Integration framework of express delivery network management model. *Source* The Author

can yield high profits to use, leading to the mutual benefit of resource utilization, and helping to a certain extent the express delivery industry to implement network management innovation.

10.4.1.2 Task Integration

The ultimate goal of the Express Delivery Network management model is to create value for customers. The industry needs to integrate tasks, maximize the utilization of specific logistics facilities within the logistics network, and properly allocate them to target customers. Task integration essentially involves providing customers with logistics facilities services that match the parameter values, primarily the network topology characteristics. From the perspective of network theory, a comprehensive logistics service chain is established, comprised of a widely distributed logistics center and its extensive connectivity, which creates synergistic opportunities for the customer (Dong, 2011). This section enables the logistics facilities that best suit the needs of the customer to be allocated through the "connected" Express Delivery Network management model.

10.4.1.3 Function Integration

Function integration primarily relies on the immediate strategy of "optimal interconnection" and "long-distance interconnection" within the express delivery network to share information, integrating the functions of order processing, transportation, and dispatching, and reducing dysfunction and material consumption to the greatest extent possible. Among them, "preferential interconnection" refers to the collaborative agreement among express delivery enterprises that have good overall operational efficiency and ratings. Enterprises benefit from each other's strengths to achieve a mutually beneficial outcome. The interconnection of the network can allow for mutual support within the express delivery network, thus drastically reducing the uncertainty factors when individual nodes make decisions.

10.4.2 Human Resource Management Innovation

10.4.2.1 Developing an Information Technology Team

Many businesses operating through franchise or alliance models have complicated structures and numerous branches, yet demonstrate lax management. They have their own systems of management that are not maintained adequately, and they have yet to construct a genuine information technology team. On the one hand, customer relationship management departments are unable to share users' information as it is input into the system by the employees of each department and there is a lack of connection between the systems of various departments, preventing resource sharing. This makes it difficult for express delivery companies to manage the customer relationship as the systems of each department cannot form a unified whole; on the other hand, when an enterprise analyzes the stored users' data, which may exist in multiple systems, it results in a lengthy query of a program loop, an inconsistent customer information query, an increased difficulty of data mining and a reduced effectiveness of data measurement. This then leads to companies being unable to accurately meet customer needs. Therefore, in order to share customer information resources among the various departments of the company, the various departments of the express delivery company must coordinate and establish a centralized and unified data management center (Zhao, 2014).

10.4.2.2 Improving the Quality of Employees

The express delivery industry is emphasizing a people-oriented concept and focuses on the selection and development of employees. Appropriate and diverse training programs have been created, such as lectures from internal and external sources and outward-bound training, for different functional departments and for employees of different levels. Additionally, the "old staff mentoring new employee" work method can be implemented to help new employees better understand the company's operations and workflow, thereby increasing the quality of employees' work and improving the brand service image.

10.4.3 Relying on the Big Data Platform

10.4.3.1 Building an Information Platform

In the big data environment, the overall conditions and circumstances for innovation are intricate and varied. The express delivery industry will confront further innovative options, techniques, and pathways to implement the management innovation of the express delivery network. Simultaneously, the selection of pathway and of these management innovation approaches will alter according to the changes in the external environment and conditions of the express delivery industry. Subsequently, due to the transformations in the external environment of the express delivery industry, it is determined that the management innovation of the express delivery industry has several significant differences between the big data environment and the traditional mode, as depicted in Table 10.1.

The information platform can be built by harnessing the potential of Big Data. The scope of express delivery network management primarily encompasses information, logistics, capital, and express delivery outlets management, etc. As the express delivery and logistics industry has seen rapid growth, the advantages of network technology have enabled the streamlining of express delivery management operational procedures, reducing transportation time and improving fault tolerance rate. As the express delivery industry moves towards informatization, automation, and networking, the express delivery network management can possess powerful information processing and transmission capacities, allowing for real-time monitoring of the goods in transit. Consequently, the management of the express delivery network has been enhanced.

Table 10.1 Management innovation differences between traditional models and big data environments

Comparative aspects	Management innovation of traditional express delivery network	Express delivery network management innovation in the big data
Drivers	Driven by problems	Innovation Issues and Data Driven
Environment and conditions	Resources within the organization	Big data within organizational ecology
Innovation pathways	Top-down	Top-down, bottom-up, network-based collaborative
Innovation approaches	Delphi algorithm, decision tree method	Cloud computing, data mining, and statistical analysis
Key success factors	Expert experience and understanding	Results of data analysis and interpretation

Source The Author

10.4.3.2 Optimizing the Operational Model

Nowadays, most cross-border enterprises in developed western countries have realized the network management model of "one flow, three networks". Among them, the one flow mainly refers to the information flow of customer orders. The three networks refer to the environmental resource network, the customer resource network and the information technology network. China's express delivery enterprises can refer to the operation mode of the "one flow, three networks", based on various information resources in the current environment, accurately positioning, further optimizing their operation mode, and providing more personalized services for different customers (Ren, 2018). The service content enables express delivery enterprises to serve customers at no distance, and at the same time, it can also reduce the waste of corporate resources to a certain extent.

10.5 Summary

Initially, new management concepts and reforms are proposed to address the issues associated with different express delivery network management models in the chapter. These reforms can be broken down into four main aspects: management mechanism reform, network system reform, transportation management reform and service reform, which will help to facilitate the sustainable development of express delivery enterprises. Subsequently, based on the analysis of the express delivery network from the preceding article, a framework mode of China's express delivery network management innovation is developed. Finally, combining the reform methodology of the management model from Chapter 9 and the framework mode of express delivery network management innovation from the third section of this chapter, an innovation strategy of express delivery network management is proposed.

References

Cao, R. Z. (2006). Enterprise eco-management: A new management paradigm, *Digest of Management Science*, (01), 27–28.

Chang, Y. P. (2013). Exploration on innovative management mode of characteristic literature information resources in the library of East China University of Political Science and Law. *Research on Library & Information Work of Shanghai Colleges & Universities, 23*(01), 1–3.

Chen, H. Z. (2017). Current situation and innovation optimization of financial management in the express delivery industry. *Money China,* (23), 90–91.

Dong, J. (2011). *Research on management mode of enterprise group logistics networks* [Doctoral dissertation, Beijing Jiaotong University].

Feng, J., Liu, J. N., & Xu, L. (2019). Research on the integration of warehousing and distribution of E-commerce logistics under the background of 'Internet+'. *Logistics Engineering and Management, 41*(07), 83–84.

Han, L. H., & Wei, M. Z. (2019). Research on innovation of information resource management model in big data environment. *Information Science, 37*(08), 158–162.
Jing, S. W., & Niu, Z. W. (2014). Definition of the concept and content of enterprise management innovation, *Chinese Journal of Management Science, 22*(S1), 654–658.
Li, F. (2014). On business model change in big data time. *China Opening Journal,* (06), 107–109.
Liu, Y. Q. (2013). Study on the models and operation management of higher education information resources sharing abroad. *Library & Information,* (02), 118–123.
Qi, D. (2019). 'Internet + Agriculture': Research on optimization of E-commerce logistics mode of agricultural products. *Agricultural Economy,* (01), 139–140.
Qin, Z. (2016). *The research on business model innovation of Chinese express enterprises in the context of new normal* [Master dissertation, Yunnan University].
Ren, P. (2018). Research on synergy between express delivery and E-commerce industry chain in big data era. *Economic Practice,* (18), 219.
Tian, T. (2016). Huawei's idea innovation and system innovation. *Enterprise Management,,* (03), 6–11.
Wang, D. H. (2011). Evolution trend of theory and practice of human nature assumption and management. *Enterprise Civilization,* (06), 62–64.
Wang, F. (2015). Current situation and innovation optimization of financial management in the express delivery industry. *Money China,* (11), 282.
Wang, Q. W., Zhu, H., & Wang, X. Y. (2017). Research trends and prospects of brand ecology theory. *Enterprise Economy, 36*(03), 14–22.
Xu, Q. (2016). Discussion on direct operation mode of logistics express (Kuayue Express Group). *Modern Business,* (06), 123–124.
Yu, S. Y., Wang, J. D., & Tong, N. N. (2016). Innovation of government information service under big data environment: Research Status and Development Countermeasures. *E-Government,* (01), 26–32.
Yu, Z. H., & Zhang, C. (2014). A comparative study on the business models of express delivery in China. *China Market,* (49), 16–17.
Zeng, X. G. (2013). *The research on the management innovation of smart phone industry in mobile internet era* [Doctoral dissertation, Wuhan University].
Zhang, D. Y., & Zheng, L. (2019). Selection of logistics distribution strategy of retail terminal under the background of new retail—From the perspective of price service. *Modern Business Trade Industry, 40*(28), 32–35.
Zhao, J. (2014) *Research on customer relationship management of international express company* [Master dissertation, Capital University of Economics and Business].

Chapter 11
Conclusions and Outlook

11.1 Main Work and Conclusions

This research focuses on the operational mechanism of the express delivery industry, the structure and optimization methods of the express delivery network, the operational models of express delivery enterprises, etc., intending to explicate the innovative management system of the express delivery network. At the theoretical research level, the concepts of the express and express delivery network are clearly defined. The connotations of network optimization theory, operational mechanism concept, and mode change theory in express delivery network management are analyzed. The paper systematically reviews the process of the express delivery industry in China and other countries. It also investigates the issues and current advantages of China's express delivery companies in the development process and examines the current state of global express delivery companies and China's state-owned and private enterprises. This research sorts out the structure models and characteristics of the express delivery network, constructs a express delivery network topology, and further analyzes the characteristics of express delivery traffic flow within the dimensions of vehicle transportation, time, and space. Additionally, the research examines the operation theory and mechanism of express delivery network as well as the management mode of the express delivery network. In terms of empirical research, Ant Colony Optimization (ACO) and Simulated Annealing (SA) are both used to optimize the express delivery network and to explore a more effective optimization method for the express delivery network. Moreover, to clarify the extent of resource optimization and utilization during the express delivery network operation, three-stage Data Envelopment Analysis (DEA) model is used to analyze the operational efficiency of the express delivery network in each region, using panel data from 2011 – 2017. This research draws the following conclusions.

P. Li and J. Wu, *Research on Innovation of Express Delivery Network Management*,
https://doi.org/10.1007/978-981-99-8982-9_11

(1) The Express delivery network topology is a hybrid of Fully-connected network and Hub-and-spoke network. Starting from the regional structure of the express delivery network, the paper analyzes the composition characteristics of express delivery network infrastructure through five parts: regional transit centers, city transit centers, service outlets, transportation routes (networks) and customers, and reveals the uneven regional distribution of infrastructure of the express delivery network. Furthermore, analyzing the characteristic of express delivery traffic flow, it exhibits a "three-stage" distribution characteristic with a rapid increase year on year on an annual time scale.. Moreover, the express traffic flow varies significantly among regions, with a few regions carrying the majority of it. This difference is attributed to the structural characteristics of the Express delivery network and the disparate level of regional economic development.
(2) By sorting out the contents and models of various intelligent algorithms, the paper found that there are great differences. Ant Colony Optimization (ACO) and Simulated Annealing (SA) algorithms are used for optimization based on the dual objectives of the shortest route and the lowest carbon emissions. The optimization process demonstrates the strong robustness of ACO and the fast-solving speed of SA. The effectiveness of these models and algorithms is proved by selecting the postal calculations in the surrounding areas of Shaanxi Province.
(3) Regional express delivery industry and network development are uneven, with a gradient that is consistent with regional division boundaries and a tendency towards intra-regional concentration. The paper compares and analyzes four classic evaluation models: Principal Component Analysis (PCA), Structural Equation Modeling (SEM), Data Envelopment Analysis (DEA), and Factor Analysis (FA). and FA was selected to evaluate the development of the express delivery network. Then, the authors selected thirteen evaluation indicators from four aspects (macroeconomic environment indicators, system input indicators, industrial development indicators and informatization indicators) for empirical research. It was found that the economic condition is the primary determinant of the regional logistics and express delivery industry's development., while system input is the secondary factor. In addition, the development gradient of the regional express delivery industry is in accordance with the regional division. In the first gradient, nine provinces are located in Eastern China. The second gradient comprises four provinces located in Central China and three in Western China. And the third gradient consists of nine provinces located in Western China. This consistency is also an indirect indication that the economic condition is the primary determinant of the regional logistics and express delivery industry's development.. Nine out of eleven provinces and cities in eastern China are among the top eleven and occupy the top seven in the development of the express delivery industry, showing a regional concentration trend. In contrast, the overall development in the central and western parts of China is lower, indicating that the development of the express delivery industry is affected by scale effects and driving effects.

(4) The operational mechanisms of express delivery networks consist of motivation, conduction, distribution, distribution, organizational management, and guarantee mechanisms. The motivation mechanism includes two components, profit-driven and cost mechanisms, which together stimulate the establishment and operation of the express delivery network. The conduction mechanism has three components: transportation management, warehousing management, and packaging management, which reflect the operational efficiency of the express delivery network. The distribution mechanism consists of four parts: self-operated distribution, third-party distribution, fourth-party distribution, and joint distribution; this is the core link of express operations and is closest to the final customer, impacting user experience. The organization and management mechanism is an organizational guarantee for the development of the express delivery network. The guarantee mechanism constrains and regularizes the express delivery network operation to serve for the express delivery network operation.

(5) The operational efficiency of express delivery networks is highly variable and is still in need of improvement. A three-stage Data Envelopment Analysis (DEA) method was utilized to analyze the operational efficiency of China's express delivery industry. The first step was the establishment of an evaluation index for the operational efficiency of express delivery networks. Using the Banker, Charnes and Cooper (BCC) model and cross-sectional data, the operational efficiency of the express delivery industry of 31 provinces in China from 2011 to 2017 was measured. It found that the overall operational efficiency of the express delivery industry was relatively low and that the regional economic development was not a determinant for efficiency, while the management had a profound impact on the operational efficiency of express delivery networks. An SFA model was used to select the panel data from 2011 to 2017, eliminating environmental variables and random errors, and making adjustments to input–output variables. The adjusted input–output variables were then calculated by DEA, making the evaluation results more reasonable and accurate. The results showed an increase in the average pure technical efficiency, but a slight decrease in both the scale efficiency and comprehensive technical efficiency. The different efficiency values showed dynamic changes with an unstable pattern of increases and decreases according to the annual data.

(6) The paper explores the reform of express delivery network management modes by evaluating the existing management modes of express delivery enterprises. Analysis reveals that the current operating modes in China are mainly composed of the direct, franchise, and self-operated modes of operation. Considering the status quo of the management modes of express delivery enterprises, this study examines the relevant influencing factors that affect the reform of express delivery network management modes, including market factors, price factors, logistics facilities, the selection of distribution outlets, and policy factors. To address the problems inherent in the current management modes, the management mechanisms, network systems, transportation management, and services

reforms are proposed. To some extent, it provides a new example for express delivery enterprises in terms of network management reform.

(7) The current situation of management innovation of China's express delivery enterprises is analyzed from two perspectives: the national environment of China and the industryl situation. At present, the main obstacles to the healthy development of China's express delivery industry have been identified in five areas, including inadequate regulations and policies, irrational management methods, lack of frontier technology and infrastructure, and inadequate training. Therefore, it is necessary to innovate the management mode of express delivery networks. Theoretically, it is essential to clarify the concept and content of the management innovation system, construct a national ecosystem for express delivery network management innovation, and propose a new management mode of express delivery networks, which is referred to as the "Multi-ecosystem Management mode". It consists of three components: a management mode of the platform of IoTs + big data, network data mining + information resource optimal management mode, and network system framework + customer core personalized service management mode. Moreover, specific innovation strategies are provided for connecting the logistics network, innovating human resource management, and uniting the Big Data platform, offering a reference for the management innovation of express enterprises.

11.2 Future Work and Outlook

The management of express delivery networks is a relatively new research topic, and relevant domestic research is limited and not yet fully developed. This research explores the theoretical mechanisms and empirical analysis and constructs a relatively comprehensive theoretical analysis framework for the innovation of express delivery networks management modes. Nevertheless, due to the limited academic capabilities and energy, there are still shortcomings in this research, and further improvement and exploration are needed in the following areas.

(1) Through the empirical analysis, the number of indicators and samples selected was limited based on the consideration of data availability. The next step should be to expand the scope of research data, focusing on two aspects: firstly, to improve the empirical indicator system by continuing to seek representative indicator variables to make indicators more comprehensive and improve the accuracy of empirical results. Secondly, more samples need to be selected to make the empirical results more convincing and stable.

(2) In this book, the operational efficiency of the express delivery network in each province is studied at the macro level, and the reasons for the differences in the operational efficiency are analyzed. However, the specific influencing factors need to be further explored. Establishing an operational efficiency evaluation system for the express delivery industry will identify factors contributing to operational efficiency in different regions and refine the evaluation system.

(3) This research studies express delivery network management from three aspects: network optimization, operational mechanism and mode reform, and proposes innovative connotations and management concepts from the technical and management levels. Further research is needed to assess the applicability of the network optimization method demonstrated in this paper to all regions. The paper proposes the direction of management mode reform and a multi-ecosystem express delivery network management mode. As the default assumption of "all express delivery products are non-different" in the research process, it does not conform with the reality of the express market. Thus, innovation of the management mode of the express market of multi-type express delivery products requires further research.

Postscript

This book has been produced as a result of the National Social Science Foundation of China's post-funded project, "Research on Innovation of Express delivery network Management- Network Optimization, Operation Mechanism and Model Reform" (18FGL022), and is a culmination of my team's and my own dedication to "Express Delivery Network" research over the past few years. After much effort and study, the book was finally completed. To gain accurate information and data concerning express delivery network construction, we visited major express companies, such as China Post, SF Express and ZTO Express, and conducted thorough field research. We also gained an understanding of the operation status and management mode of the express delivery network, thus ensuring the accuracy of information resources and the rationality of analysis.

Now, this book has been completed. I am immensely grateful to all the research members who have worked diligently for it, which brings a sense of accomplishment and joy.

First of all, I would like to extend my gratitude to all of the team members who participated in the book. Their dedication and hard-work, which included sacrificing vacation time, conducting field visits to obtain materials, and researching overtime, were integral to the smooth progression and success of the project. Their dedication and commitment should be an example for everyone to admire and respect. Secondly, I would like to thank the relevant companies for their cooperation and assistance in the research. Their participation made the project results more realistic. Thirdly, I want to thank my students for their active participation, which enabled us to overcome challenges and advance our academic exploration, thus allowing the project to progress rapidly and succeed after perseverance. Additionally, I would like to thank the experts who provided valuable opinions during the evaluation process of the National Social Science Fund project. Their insight enabled us to make the book more perfect through continuous revisions. Finally, I want to express my gratitude to many scholars who have inspired the research process and the writing of this book. Unfortunately, I cannot thank them all individually, so please forgive me.

P. Li and J. Wu, *Research on Innovation of Express Delivery Network Management*,
https://doi.org/10.1007/978-981-99-8982-9

Last but not least, I would like to express my gratitude to my family. The attentive care provided by my parents, the constant patience and support from my wife, and the healthy growth and consideration displayed by my daughter have all served as invaluable sources of inspiration and motivation for me to persist in my scientific endeavors.

References

Amagai, Y., Shoji, K., & Toyama, F. (1983). Optimization by simulated annealing. *Science, 42*(3), 671–680.

Armacost, A. P., Barnhart, C., & Ware, K. A. (2002). Composite variable formulations for express shipment service network design. *Transportation Science, 36*(1), 1–20.

Armacost, A. P., Barnhart, C., Ware K. A., & Wilson, A. M. (2004). UPS optimizes its air network. *Interfaces, 34*(1), 15–25.

Atif, Y. A. (2004). A distributed and interoperable object-oriented support for safe e-commerce transactions. *Information and Software Technology, 46*(3), 149–164.

Bai, M. G., & Zhu, J. F. (2006). Comparative study on fully-connected and hub-and-spoke airline networks. *Systems Engineering Theory and Practice, 26*(9), 113–117.

Barnhart, C., & Schneur, R. R. (1996). Air network design for express shipment service. *Operations Research, 44*(6), 852–863.

Brueckener, J. K., & Zhang, Y. (2001). A model of scheduling in airline networks: how a hub and-spoke system affects flight frequency, fares an welfare. *Journal of Transport Economics and Policy, 35*(2), 195–222.

Bu, X. H., Li, L., & Wang, H. Z. (2012). Study on the influence of informatization development on service level of express delivery industry. *China CIO News*, 6, 117–118.

Buchmiller, J. L., & Connell, C. L. (2003). Innovation engine portal method and system, US, US 20030187706 A1.

Cai, B., Yao, Z. Q., & Zhang, Q. (2008). The application of principal component analysis in the developmental status of economy of Jiangsu province. *Mathematics in Practice and Theory*, (11), 7–16.

Cao, R. Z. (2006). Enterprise eco-management: a new management paradigm. *Digest of Management Science*, (1), 27–28.

Chang, Y. P. (2013). Exploration on innovative management mode of characteristic literature information resources in the library of East China University of Political Science and Law. *Research on Library & Information Work of Shanghai Colleges & Universities*, *23*(1), 1–3.

Charnes, A., Cooper, W. W., & Rhodes, E. (1978). Measuring the efficiency of decision making units. *European Journal of Operational Research*, *2*(6), 429–444.

Chen, B. (2016). Research on the interactive relationship between e-commerce and express delivery industry: dynamic empirical analysis based on VAR model. *Journal of Fujian Normal University (Philosophy and Social Science Edition)*, 1, 63–69+82.

Chen, H. Z. (2017). Current situation and innovation optimization of financial management in the express delivery industry. *Money China*, (23), 90–91.

Chen, L. (2015). *Information technology progress and competitive strategy choice of private express delivery enterprises*. Beijing: China Social Science Press.

P. Li and J. Wu, *Research on Innovation of Express Delivery Network Management*,
https://doi.org/10.1007/978-981-99-8982-9

Chen, L., Wang, H. B., Sun, H. T., & Liu, C. X. (2017). The Return of MOOCs in China and the direction of reform of teaching service mode in colleges and universities. *China Educational Technology*, (8),1–6+12.

Chen, Y., Du, J., Sherman, H. D., & Zhu, J. (2010). DEA model with shared resources and efficiency decomposition. *European Journal of Operational Research, 207*(1), 339–349.

Chen, Z. X., & Du, X. J. (2016). Research on new profit model of express delivery enterprises under Internet+. *Pioneering with Science & Technology Monthly, 29*(21), 24–25.

Chen, Q. (2014). Discussion on development mode of regional private express industry: based on sample enterprises from Central-Zhejiang City Group. *Logistics Technology, 33*(1),245–247+317.

Cheng, H. (2002). The impact of the internet on business models. *Commercial Research*, (3), 140–141.

Chestler, L. (1985). Overnight air express: spatial pattern, competition and the future in small package delivery services. *Transportation Quarterly, 39*(1), 59–71.

Cormen, T. H. (2006). *Introduction to algorithms.* Beijing: China Machine Press.

Cui, X. Y. (2016). Analysis of profit model of logistics enterprises based on internet of things. *Journal of Commercial Economics, 28*, 36–37.

Cunha, C. B., & Silva, M. R. (2007). A genetic algorithm for the problem of configuring a hub-and-spoke network for a LTL trucking company in Brazil. *European Journal of Operational Research, 179*(3), 747–758.

Dai, T., & Huo, J. Z. (2010). Research on express shipment network design based on non-hub network. *Computer Engineering Application, 46*(5), 5.

Dasgupta, D., & Michalewicz, Z. (1997). *Evolutionary algorithms in engineering applications . DBLP.*

De Camargo, R. S., Miranda Jr, G., Ferreira, R. P. M., & Luna, H. P. (2009). Multiple allocation hub-and-spoke network design under hub congestion. *Computers & Operations Research, 36*(12), 3097–3106.

Dijkstra, E. (1959). A note on two problems with graphs. *Numerische Mathematik*, (1), 269–271

Ding, J. C., Tong, W. J., Huang, Z. X. (2008). Problems and innovation of cooperative mechanism between higher vocational college and enterprise. *Research in Educational Development*, (17), 67–70.

Dong, J. (2011). *Research on management mode of enterprise group logistics networks* [Doctoral dissertation, Beijing Jiaotong University].

Dorigo, M., Birattari, M., & Stutzle, T. (2006). Ant colony optimization. *Computational Intelligence Magazine IEEE, 1*(4), 28–39.

Du, J., & Huo, J. Z. (2014). DEA-based evaluation on city innovation in China. *Chinese Journal of Management Science, 22*(6), 85–93.

Duan, S. L. (2015). Empirical study on influence factors of Chinese express industry development. *Logistics Engineering and Management, 37*(1), 171–173.

ErnÝ, V. (1985). Thermodynamical approach to the traveling salesman problem: an efficient simulation algorithm. *Journal of Optimization Theory and Applications*, 45(1), 41–51.

Fan, C. H. (2007). *Research on CRE logistics network planning based on Hub and Spoke Theory* [Master dissertation, Beijing Jiaotong University].

Fan, J. H., & Wang, X. F. (2006). Vehicle routing optimization problem based on immune algorithm. *Computer Engineering and Applications*, (4), 210–212+217.

Feng, J., Liu, J. N., & Xu, L. (2019). Research on the integration of warehousing and distribution of E-commerce logistics under the background of 'Internet+'. *Logistics Engineering and Management, 41*(7), 83–84.

Feng, X. Z., He, Z. L. (2010). Research on the operating mechanism of the organizational knowledge innovation system. *Information Studies: Theory & Application, 33*(1), 15–19.

Feng, H. (2007). *The research on Chinese regional industrial competitiveness: theory and empirical analysis* [Master dissertation, Jilin University].

Fisher, M. L. (1994). Optimal solution of vehicle routing problems using minimum KK-trees. *Operations Research, 42*(4), 626–642.
Floyd, R. W. (1962). Algorithm 97 shortest path. *Communications of the ACM, 5*(6), 345.
Fu, Y. Y. (2009). *The design of supply chain networks and fuzzy multi-objective optimization based-on fuzzy theory* [Doctoral dissertation, Zhejiang University].
Gamboa, A. L. S., Filipe, E. J. M., Brogueira, P. (1995). Coordination in software development. *Communications of the ACM, 38*(3), 69–81.
Gao, B., & Tao, B. G. (2013). *Introduction to express service.* Beijing: Posts & Telecom Press.
Ge, X., Yu, B., & Jin, Z. H. (2013). Optimization of regional express network based on time threshold and tariff discounts. *Journal of Dalian Maritime University, 39*(2), 73–77+81.
Gu, L. P. (2018). Data management services in the transition of research model: an approach of implementing open access, open data and open science. *Journal of Library Science in China, 44*(6), 43–58.
Guo, D. H. (2016). *Internet New Business Model.* Beijing: China Machine Press,.
Guo, H. (2013). *Research on development strategy of express delivery business of China postal express logistics company* [Master dissertation, Beijing Jiaotong University].
Halman, J. I. M., & Keizer, J. A. (1994). Diagnosing risks in product-innovation projects. *International Journal of Project Management, 12*(2), 75–80.
Han, L. H., & Wei, M. Z. (2019). Research on innovation of information resource management model in big data environment. *Information Science*, 37(8), 158–162.
He, B. Q., Li, K. P., Cheng, X. X. (2019). The research on the optimization of courier companies "Last-Mile" express pickup and delivery process. *Operations Research and Management Science*, (1), 27–34.
Hedlund, G. (1994). A model of knowledge management and the n-form corporation. *Strategic Management Journal*, 15(S2), 73–90.
Holland, J. H. (1975). Adaptation in Natural and Artificial Systems. Ann Arbor: MI University of Michigan Press.
Holland, J. H. (1992). Genetic algorithms. *Scientific American*, 267, 66–72.
Horner, M. W., & O'Kelly, M. E. (2001). Embedding economies of scale concepts for hub network design. *Journal of Transport Geography, 9*(4), 255–265.
Hu, X. D., & Zhang, H. S. (2001). Applying hierarchy model to multiobjective decision making on postal transportation operations. *Systems Engineering and Electronics*, (9), 74–77.
Huang, J. H. (2009). Analysis on the complexity and robustness of express distribution network—a case study of an express enterprise. *Journal of Southwest Jiaotong University (Social Sciences), 10*(6), 98–102.
Huang, S. Y. (2010). Interpret the management specifications of postal dispatch. *China Post News*, 2010–02–06(008).
Jeong, S. J., Lee. C. G., & Bookbinder, J. H. (2007). The European freight railway system as a hub-and-spoke network. *Transportation Research Part A: Policy and Practice, 41*(6), 523–536.
Ji, C. X., Liu, B. D., & Zhuo, X. Y. (2015). Research on innovation and comparison express delivery terminal operating model on campus in the e-commerce environment. *Business*, 39, 263–264.
Jia, J. K. (2009). Analysis of the relationship between principal component and factor analysis. *Knowledge Economy*, (7), 168–169.
Jiang, B., & Li, J. (2015). Domestic market size and industrial upgrading of China's express industry. *China Business and Market, 29*(1), 77–83.
Jiang, J. C., Wu, L. X., Yang, Y. Z., & Li, Z. F. (2014). Self-adaptive algorithm for network maximum flow problem: SAPR algorithm. *Application Research of Computers, 31*(10), 2969–2973.
Jiang, L. L. (2018). *A spatial analysis of the express delivery network of "China Railway Express"* [Master dissertation, Shanghai Normal University].
Jing, S. W., & Niu, Z. W. (2014). Definition of the concept and content of enterprise management innovation. *Chinese Journal of Management Science, 22*(S1), 654–658.
Kennedy, J., & Eberhart, R. (1995). *Particle Swarm Optimization. Icnn95-international Conference on Neural Networks* [Conference]. IEEE, New York, USA.

Kirkpatrick, S., Gelatt, C. D., & Vecchi, M. P. (1983). Optimization by simulated Annealing. *Science*, 220, 671–689.

Kleppe, A. G. (2017). First European workshop on composition of model transformations—CMT 2006. *Journal of Zoological Systematics & Evolutionary Research, 33*(3–4), 42–47.

Kohl, N., & Madsen, O. (1997). An Optimization Algorithm for the vehicle routing problem with time windows based on Lagrangian relaxation. *Operations Research, 45*(3), 395–406.

Kong, F. Z. (2017). Research on enterprise business model innovation path under the background of internet plus—a case study of manufacturing enterprises. *China Journal of Commerce, 20*,133–135.

Koza, J. R. (1992). *Genetic programming: on the programming of computers by means of natural selection*. Cambridge: MIT Press.

Kuby, M. J., & Gray, R. G. (1993). The hub network design problem with stopovers and feeders: the case of federal express. *Transportation Research Part A: Policy and Practice, 27*(1), 1–12.

Lee, H. J., Ko, H. J., & Sohn, Y. H., et al. (2008). *Profit-based network design in express courier services*. International Conference on Innovative Computing Information & Control [Conference]. IEEE, New York, United States.

Li, F. (2014). On business model change in big data time. *China Opening Journal*, (6), 107–109.

Li, H. (2001). Carry reform through to the end. *Enterprise Management*, 10, 1.

Li, H., Qin, Q. L., & Fang, D. L. (2014). Case study of bonded logistics system in interconnection. *Economic Review Journal, 3*, 71–74.

Li, J. (2017). Research on the operation mode of modern logistics park in Linying, Henan Province. *Journal of Henan College of Finance & Taxation, 1*, 63–66.

Li, L. (2012). *Research on national port price mechanism based on DEA Model* [Master dissertation, Shanghai Jiao Tong University].

Li, L., & Ding, Y. Z. (2012). Optimization of hub location and allocation for hub-spoke express delivery network. *Journal of Shanghai Maritime University, 33*(2), 33–39+75.

Li, L., & Yang, H. Z. (2018). Impacts of operational mechanism of network platform on new service development performance of content provider. *Science Research Management, 39*(1), 18–25.

Li, L., Zhang, M., & Wang, J. (2013). Research on improving the price formation mechanism of express industry in China—a brief discussion on the application of step method in express pricing. *Price: Theory & Practice*, (6), 54–55.

Li, M., & Liu, Y. (2011). Comparative study of courier industry in eastern and western regions. *Value Engineering, 30*(4), 24–25.

Li, N., Zou, T., & Sun, D. B. (2004). Particle swarm optimization for vehicle routing problem. *Journal of Systems Engineering*, (6), 596–600.

Li, P. (2008). New algorithm based on simulate anneal and hopfield neural for solving TSP. *Science Technology and Engineering, 8*(14), 3937–3939.

Li, R. D. (2018). A study on the business model of rural express delivery enterprise. *Agricultural Science-Technology and Information,* (14), 68–71.

Li, R., Liu, L., & Zheng, X. N. (2017). Supplier-facilitated inventory transshipments: an innovative approach based on firm practice. *Journal of Management Science, 30*(6), 92–103.

Li, W. X. (2019). *A study on the reform of teaching and research group's professional autonomy* [Doctoral dissertation, East China Normal University].

Li, Y. (2006). *Research on hub and spoke network theory and its application*. [Doctoral Dissertation, Fudan University].

Li, Z. H. (2015). Research on the selection of SF Express operation mode. *Logistics Engineering and Management, 37*(7), 47–50.

Li, Z. W., & Zhang, X. M. (2006). Study on the optimization of logistics distribution network based on dynamic sweep and Ant Algorithm. *Journal of Engineering Management*, (4), 9–12.

Liao, H. M., Liao, X. H., & Chen, H. K. (2015). Study on highway flood fuzzy probability comprehensive evaluation model based on method of factor analysis. *China Safety Science Journal, 25*(2), 141–146.

Lin, C. C., & Lin, J. S. J. (2007). The multistage stochastic integer load planning problem. *Transportation Research Part E: Logistics and Transportation Review, 43*(2), 143–156.
Lin, F. (2017). Drive, transformation and innovation: SMG's institutional reform concept and path. *TV Research*, (7), 15–17.
Lin, S. S. (2016). Research on express industry in the e-commerce environment—in the case of SF. *Logistics Engineering and Management, 38*(8), 81–82.
Lin, T., & Xie, X. C. (2017). Elements, structure of express delivery network and their spatial pattern in Shanghai. *Transport Research, 3*(5), 9–49.
Liu, J. B., & Pan Gao, Study of the GEM listed companies' performance evaluation in China based on BBC model. *China Soft Science*, (S1), 263–270.
Liu, J. M., Ma, S. F., & Ma, S. Q. (2011). Computation method of the dynamic shortest path based on improved-Dijkstra algorithm. *System Engineering—Theory & Practice*, 31(6), 1153–1157.
Liu, P. (2016). *Research on county location model of rural express delivery business outlets* [Master dissertation, Chang'an University].
Liu, Y. Q. (2013). Study on the models and operation management of higher education information resources sharing abroad. *Library & Information*, (2), 118–123.
Lu, H. S., Gao, X. D., & Wu, S. (2003). An algorithm for maximum profit flows. *Mathematics in Practice and Theory, 33*(5), 43–48.
Lu, S. (2010). Inspiration from the development of German express industry. *China Logistics and Purchasing, 2*, 6–7.
Lu, W. T. (2016) Regional express network optimization based on time threshold. *Journal of Wenzhou University (Natural Science), 37*(2), 46–54.
Lu, X. W., & Yuan, Y. J. (2008). Process of industrial evolution based on enterprise organization reform. *Journal of Dalian University of Technology (Social Sciences), 29*(4), 31–35.
Ministry of Public Security of State Post Bureau, Ministry of State Security, "Joyfully Announces Regulations on the Administration of Prohibited Items" State Post Bureau of The People's Republic of China. Retrieved from http://www.spb.gov.cn/zc/flfgjzc_1/201612/t20161216_934646.html2017.08.24.
Ni, L. L. (2010). Analysis of the characteristics of hub-and-spoke, point-to-point, and combined express networks. *Statistics and Decision*, (20), 59–61.
Ni, L. L. (2012). *Study on express operation network optimization design and competitive network equilibrium* [Doctoral dissertation, Central South University].
O'Kelly, M. E. (1987). A quadratic integer program for the location of interacting hub facilities. *European Journal of Operational Research, 32*(3), 393–404.
Osterwalder, A., Yves, P., & Chirstopher, L. T. (2005). Clarifying business models: origins, present, and future of the concept. *Communications of the Information Systems, 15*(5), 1–25.
Padberg, M., & Rinaldi, G. (1991). A Branch-and-Cut Algorithm for the resolution of Large-scale symmetric traveling salesman problems. *SIAM Review, 33*(1), 60–100.
Pels, E., Nijkamp, P., & Rietveld, P. (2000). A note on the optimality of airline networks. *Economics Letters, 69*(3), 429–434.
Postal Law of the People's Republic of China. State Post Bureau of The People's Republic of China [EB/OL]. Retrieved from http://www.spb.Gov.cn/zc/flfgjzc_1/201210/t20121026_159203.html.2017.08.24.
Qi, D. (2019). 'Internet + Agriculture': research on optimization of e-commerce logistics mode of agricultural products, *Agricultural Economy*, (1), 139–140.
Qian, S. D. (1990). *Operations research.* Beijing: Higher Education Press.
Qin, L. G., Wang, N. N., & Li, J. (2015). A DEA measurement and optimization of comprehensive efficiency of regional express delivery industry: in the case of Guangxi. *Logistics Technology*, 34(09), 108–111.
Qin, Z. (2016). *The research on business model innovation of Chinese express enterprises in the context of new normal* [Master dissertation, Yunnan University].
Rechenberg, I. (1978). *Evolution strategien. Simulations methoden in der Medizin und Biologie.* New York: Springer Berlin Heidelberg.

Ren, D. Y. (2014). Study on a new university on-campus express delivery mode under student autonomy, *Logistics Technology, 33*(11), 71–73.

Ren, M. M., & Yang, C. (2007). Optimized design of waste joint recycling logistics network. *Statistics and Decision*, (14), 34–36.

Ren, P. (2018). Research on synergy between express delivery and E-commerce industry chain in big data era. *Economic Practice*, (18), 219. Retrieved from https://baike.baidu.com/item/管理模式/10595604?fr=aladdin

SF Express website. (2018, April 5). Retrieved from http://www.sf-express.com/cn/sc/express/express_servicemainland_area/.

Smirti, M., Boubert, A., Calloud, V., & Papson, A. (2007). Modeling the logistics of FedEx International Express. *National Urban Freight Conference* [Conference]. University of California at Berkeley Institute for Transportation Studies, Long Beach, United States. Retrieved from http://www.metrans.org/nuf/2007/documents/Smirti-000.pdf.2090928.2007.

Song, H. B., & Song, Y. L. (2003). An improved genetic algorithm of vehicle routing problem with time window constrains. *Journal of Transport Information and Safety, 4*, 25–27.

Sun, H. Q., & Rong, N. N. (2018). Reform of retail enterprise management mode in e-commerce environment. *Journal of Commercial Economics, 4*, 100–102.

Sun, K., & Yang, W. M. (2014). Research on dairy transportation and distribution in low-carbon economy. *Jiangsu Agricultural Sciences, 42*(4), 392–395.

Sun, L. R. (2005). The principle and operation of Structural Equation Modeling (SEM), *Journal of Ningbo University (Educational Science Edition)*, (2), 31–34+43.

Tan, K. C., Chew, Y. H., & Lee, L. H. (2006). A hybrid multi-objective evolutionary algorithm for solving truck and trailer vehicle routing problems. *European Journal of Operational Research, 172*(3), 855–885.

Tan, W. H., & Xu, C. G. (2016). Analysis of the operation mechanism of financial companies under the guidance of coordination and control. *Finance & Accounting*, (8), 50–52.

Tan, X. (2015). *Analyzing and modeling express shipping service network based on logistics data* [Master dissertation, Zhejiang University].

The State Council of the People's Republic of China. The State Council of People's Republic of China, Services Opinions of the State Council on Promoting the Development of the Express Delivery Industry" [EB/OL].Retrieved from http://www.gov.cn/zhengce/content/2015-10/26/content_10256.htm.2017.08.24.

Tian, T. (2016a). Huawei's idea innovation and system innovation. *Enterprise Management*, (3), 6–11.

Tian, T. (2016b). Reform of contemporary agricultural economy and transformation of management mode. *Agricultural Economy, 5*, 36–38.

Tian, X. Y. (2017). Analysis of advantages and disadvantages of Jingdong logistics distribution mode and its countermeasures. *Modernization of Management, 37*(6), 92–96.

Tsoukas, H., & Chia, R. (2002). On organizational becoming: rethinking organizational change. *Organization Science, 13*(5), 567–582.

Tu, C. (2004). *Management change—carry out management innovation to the end*. Beijing: China Logistics Publishing House.

Van de Ven, A. H. (1986). Central problems in the management of innovation. *Management Science, 32*(5), 590–607.

Wang, B. Y. (2016). An analysis on regional disparities and dynamic evolution of express industry development in China. *China Business and Market, 30*(2), 36–44.

Wang, D. H. (2011). Evolution trend of theory and practice of human nature assumption and management. *Enterprise Civilization*, (6), 62–64.

Wang, F. (2015). Current situation and innovation optimization of financial management in the express delivery industry. *Money China*, (11), 282.

Wang, F. F. (2015). *A comparative study on the development mode of domestic and international express delivery industry* [Master dissertation, Hebei University of Technology].

Wang, F. M. (2017). On the current problems and opportunities in the development of China's express delivery industry. *Ability and Wisdom,* (16), 265.

Wang, J. (2015). *Optimization research of air express backbone network based on scale economies* [Master dissertation, Beijing Jiaotong University].

Wang, J. (2018). Classification and influencing factors of terminal delivery service mode of urban express delivery. *Journal of Commercial Economics, 8,* 81–85.

Wang, L., Chen, Y. Z., Fan, Y., & Xu, Z. W. (2016). Analysis of the operation mechanism of the national postal industry and its efficiency: based on the evaluation of the two-stage DEA. *Journal of Beijing Jiaotong University (Social Sciences Edition), 15*(3), 104–113.

Wang, M. Q., Liang, L., & Li, Y. J. (2009). Fuzzy super-efficiency DEA model. *Chinese Journal of Management Science, 17*(2), 117–124.

Wang, Q. W., Zhu, H., & Wang, X. Y. (2017). Research trends and prospects of brand ecology theory. *Enterprise Economy, 36*(3), 14–22.

Wang, W., & He, G. W. (2008). Research on the community criterion and the operational mechanism of rural saving and credit cooperatives. *Issues in Agricultural Economy, 9,* 23–28+110.

Wang, Y. (2016). *Enterprise reform in the internet era. Beijing:* Tsinghua University Press.

Wang, Y., & Cheng, F. X. (2012). Research and effect evaluation of resource marketing strategy model based on structural equation. *Contemporary Economics,* (12), 134–135.

Wang, Z. T. (2001). *Enterprise informatization and management reform.* Beijing: China Renmin University Press.

Wang, Z. Q., Zhu, H., & Qin, M. X. (2014). *New urbanization and transportation development—proceedings of the annual meeting and 27th symposium on urban transportation planning in China in 2013* [Conference]. Urban Transportation Planning Academic Committee of China Urban Planning Society, Beijing, China.

Wasner, M., & Zaepfel, G. (2004). An integrated multi-depot hub-location vehicle routing model for network planning of parcel service. *International Journal of Production Economics, 90*(3), 403–419.

Webster, B., & Bernhard, P. J. (2003). *A local search optimization algorithm based on natural principles of gravitation. Proceedings of the International Conference on Information and Knowledge Engineering* [Conference]. DBLP, Las Vegas, Nevada, USA.

Wei, J., & Wang, J. (2011). Research on urban logistics capacity evaluation based on DEA and BP neural network. *China Market,* (28), 21–22+25.

Weijun, L. (2014). *Discussion on 7S Standardized Management Mode of Express Enterprise Transfer Center.* State Post Bureau of The People's Republic of China; China Express Association; Conference on the Development of Chinese (International) Express Industry. State Post Bureau of The People's Republic of China; China Express Association, 8(100–107).

Wen, X. R., Jiang, X. L., Zheng, Y., Xie, X. J., Li, C. G., & Yu, G. H. (2014). Evaluation of stand volume and related factors of pinus massoniana artificial forests. *Journal of Central South University of Forestry & Technology, 34*(11), 32–36.

Wu, J. H. (2005). *Application of modern cost management methods in enterprise logistics cost management* [Master dissertation, Southwestern University of Finance and Economics].

Wu, P. (2017). Research on the development path of express delivery industry from the perspective of collaborative innovation. *Jinan Journal (Philosophy & Social Sciences), 39*(3), 25–29.

Wu, Y. T., & Lin, R. (2010). Discussion on the operation mechanism of logistics public information platform. *Fujian Transportation Science and Technology, 5,* 116–119.

Wu, Z. F., Lin, M. H., & Chen, W. (2007). Based on characteristics of regional economy of Fujian Province aquatic products cold-chain logistics development research. *Journal of Harbin University of Commerce (Social Science Edition), 1,* 100–107.

Xiang, G. P., Yu, J. H., & Huang, W. (2016). The international experience and implication of the operation mechanism of technology business accelerator. *Science & Technology Progress and Policy, 33*(20), 30–36.

Xie, F. R. (2009). *Research on efficient & robust algorithms for some problems in network optimization and its application in management* [Doctoral dissertation, Nanchang University].

Xie, J. X., Xing, W. X., & Wang, Z. B. (2009). *Network optimization*. Beijing: Tsinghua University Press.

Xie, W. T. (2009). Regional logistics and empirical study in China's provinces and cities. *Modern Business Trade Industry, 21*(06), 69–70.

Xu Z. B., Gu, T. L., & Zhao L. Z. (2005). A novel symbolic ADD Algorithm for maximum flow in networks. *Journal on Communications*, (2), 1–8.

Xu, J., Huang, D. L., & Xiao, Y. (2007). Research on the construction and operation mechanism of regional logistics information network system. *China Water Transport, 11*, 126–127.

Xu, Q. (2016). Discussion on direct operation mode of logistics express (Kuayue Express Group). *Modern Business*, (6), 123–124.

Xu, S. Q. (2016). The fuzzy comprehensive evaluation model and empirical analysis of the effect of enterprise application innovation methods. *Science and Technology Management Research, 36*(6), 79–84.

Xu, Y. D. (2003). *Research on the operation mechanism of modern agricultural science and technology park* [Master dissertation, Sichuan Agricultural University].

Xu, Z. B., Gu, T. L., & Zhao, L. Z. (2005). A novel symbolic ADD algorithm for maximum flow in networks. *Journal on Communications, 26*(2), 1–8.

Yan, X. T., & Wu, M. Q. (2009). Research on multi-objective optimization for shortest path algorithm based on GA. *Control and Decision, 24*(7), 1104–1109.

Yang, C. P., Zheng S. J., Dang Y. J., & Yang, Q. (2016) Optimization of express delivery network based on delivery timeliness and connection cost. *Systems Engineering-Theory & Practice, 36*(8), 1983–1992.

Yang, C. P., Zheng, S. J., Dang, Y. J., & Cui, J. Q. (2015). Optimization of express network based on the constraint of delivery time and node flow. *Systems Engineering, 33*(11), 53–59.

Yang, H. B. (2017). The change of science and technology management mode from the perspective of innovation ecology. *Scientific Management Research, 35*(6), 43–46.

Yang, H. R., Gu, L. Y., & Zheng, J. (1996). Organization and management of postal communication network. *Beijing University of Posts and Telecommunications Press*, 5–26.

Yang, H., Nie, Y., Zhang, H., Di, Z., & Fan, Y. (2009). Insight to the express transport network. *Computer Physics Communications, 180*(9), 1511–1515.

Yang, W. Y. (2002) On the forming method of human capital and the construction of its operational mechanism. *Modern Finance and Economics-Journal of Tianjin University of Finance and Economics, 4*, 45–50.

Yang, Y., Jin, Y. M., Sun, X. Y., & Chen, J. (2005). The source of atmospheric particulate matter in Fushun was analyzed by factor analysis. *Environmental Protection and Circular Economy*, (4), 7–9.

Yang, Z. G. (2018). On the reform of the higher education governance model in France-historical institutionalism, A double examination of institutional isomorphism theory. *Contemporary Education Sciences, 2*, 75–78.

Yang, Z. Z., & Zou, W. Q. (2011). Network optimization of RPTB express based on Ant Colony Algorithm. *Journal of Transportation Systems Engineering and Information Technology, 11*(1), 90–95.

Yao, B. H., Cui, Z. W., Xu, M. Y., & Meng, L. (2016). Research and promotion of national express business model based on O2O model. *Computer Knowledge and Technology, 10*, 285–286.

Ye, M. H., & Qiu, M. (2007). The changing management based on reality. *Business and Management Journal, 19*, 94–96.

Ye, Y. H., & Wang, L., et al. (2004). The optimal design method for the Nation's Post Network. *Journal of Industrial Engineering and Engineering Management, 2*, 39–43.

Yu, B. Q., & Du, G. W. (2013). SERVQUAL-Model-Based Fuzzy Evaluation of express service quality in online shopping environment. *Industrial Engineering Journal, 16*(2), 127–133.

Yu, S. Y., Wang, J. D., & Tong, N. N. (2016). Innovation of government information service under big data environment: research status and development countermeasures. *E-Government*, (1), 26–32.

Yu, Z. H., & Zhang, C. (2014). A comparative study on the business models of express delivery in China. *China Market,* (49), 16–17.
Yuan, C. H., Du, H., & Zhang, Y. (2016). Informatization standard system of express industry. *Journal of Beijing University of Posts and Telecommunications (Social Sciences Edition), 18*(5), 61–66+94.
Zeng, X. G. (2013). *The research on the management innovation of smart phone industry in mobile internet era* [Doctoral dissertation, Wuhan University].
Zhang, X. F., & Lin, X. F. (1998). Reform of enterprise organization mode in the era of knowledge economy. *Journal of Xi'an Jiaotong University (Social Sciences), 4,* 36–38.
Zhang, W. B. (2015). *The air express network planning study of F Company* [Master dissertation, East China University of Science and Technology].
Zhang, A. F., & Che, X. H. (2011). An analysis on the efficiency of Guangzhou technology innovation and its influence factors based on DEA. *Science and Technology Management Research, 31*(24), 20–23.
Zhang, B. Y. (2018). The general office of the state council issues the opinions on promoting the integrated development of e-commerce, delivery and logistics. *Computer & Network, 44*(5), 9.
Zhang, C., Pan, T., & Tian, R. F. (2016). Based on WeChat public platform, a study on the micro-express service mode of campus: a case study of Tianjin Agricultural College. *Education and Teaching Forum, 32,* 85–86.
Zhang, D. Q., & Wu, G. L. (2009). Optimization of floyd algorithmic for shortest path problem. *Journal of Xuchang University, 28*(2), 10–13.
Zhang, D. Y., & Zheng, L. (2019). Selection of logistics distribution strategy of retail terminal under the background of new retail—from the perspective of price service. *Modern Business Trade Industry, 40*(28), 32–35.
Zhang, F. L., & Jia, C. X. (2017). Products innovation method based on classification and importance evaluation of user needs. *Packaging Engineering, 38*(16), 87–92.
Zhang, H. B., & Nie, Y. C. (2009). Analysis on network system of China express delivery enterprises. *Logistics Technology, 28*(9), 35–37.
Zhang, J., Wu, Y. H., Liu, P., & Wang, Y. Y. (2008). Hybrid hub-and-spoke network planning of road express freight. *Journal of Shandong University (Engineering Science), 38*(5), 6–9+13.
Zhang, L. (2008). *Research on net point layout of express enterprise* [Master dissertation, Central South University].
Zhang, L. X., Zhao, Y. Q., & Pan, F. Q. (2009). Optimal path solution of road network based on Hopfield neural network. *Journal of Harbin Institute of Technology, 41*(9), 222–224.
Zhang, L., & Zhu, H. Q. (2015). Grey correction analysis of influencing factors of China's express delivery industry. *Journal of Commercial Economics, 29,* 122–124.
Zhang, L., Qi, E. S., & Chang, Q. (2014). Evaluation of management innovation methods types for manufacturing enterprises: an empirical analysis based on support vector machine. *Studies in Science of Science, 32*(11), 1747–1753+1760.
Zhang, W. B. (2015). *The air express network planning study of F Company.* [Unpublished master dissertation, East China University of Science and Technology].
Zhang, Y. X., & Jiang, H. (2017). Discussion on the reform of university teaching mode using micro-class as an opportunity. *The Party Building and Ideological Education in Schools, 6,* 35–37.
Zhang, Z. X., & Shao, B. L. (2008). Vehicle routing and scheduling problems based on improved ACA. *Journal of Highway and Transportation Research and Development, 4,* 137–140.
Zhao, G., Liu, B., & Li, Z. B. (2017). The influence of entrepreneurial behavior of incubation enterprises on entrepreneurial performance and promoting policies—a case study of Tianjin. *Enterprise Economy, 36*(1), 88–95.
Zhao, J. (2014). *Research on customer relationship management of international express company* [Master dissertation, Capital University of Economics and Business].
Zhao, J. Y., Wang Y., & Hu, D. W. (2004). Ant Colony Algorithm of distribution network planning, *Journal of Traffic and Transportation Engineering,* (3), 79–81.

Zhao, J., Zhang, J. J., & Yan, C. H. (2016). Research on hybrid hub-spoke express network decision with point-to-point direct shipment. *Chinese Journal of Management Science, 11*, 58–65.

Zhao, L. M. (2004). Study on the operation mechanism of digital publishing alliance: based on social network analysis. *Publishing Journal, 22*(6), 74–77.

Zhao, X. P., & Song, X. (2016). Application of O2O+crowdsourcing in logistics field. *Logistics Technology, 8*, 36–38.

Zhao, Y. Z., Zhang, H. F., & Wang, F. (2016). Research on the pricing mechanism reform of China's express service and relevant suggestion. *Prices Monthly,* 4, 12–17.

Zheng, S. M. (2006). *Business model reform.* Guangdong: Guangdong Economy Publishing House.

Zhou, S. F., & Liu, L. (2004). The multi goal optimization problem in network project. *Journal of East China Jiaotong University, 21*(2), 10–13.

Zhou, X. L. (2012). A discussion on express industry development under the circumstances of e-commerce rapid development. *Enterprise Economy, 31*(5), 110–113.

Zhu, C. Z., & Li, Y. L. (2013). Research on vehicle path optimization with the smallest carbon emissions. *Computer Engineering and Applications, 49*(22), 15–18.